shaping
the Next One Hundred Years

**New Methods for
Quantitative,
Long-Term Policy
Analysis**

Robert J. Lempert
Steven W. Popper
Steven C. Bankes

Prepared for

The RAND
PARDEE CENTER

This research in the public interest was supported by a generous grant from Frederick S. Pardee to develop new methods for conducting longer term global policy and improving the future human condition.

Library of Congress Cataloging-in-Publication Data

Lempert, Robert J.
    Shaping the next one hundred years : new methods for quantitative, long-term policy analysis / Robert J. Lempert, Steven W. Popper, Steven C. Bankes.
        p. cm.
    "MR-1626."
    Includes bibliographic references.
    ISBN 0-8330-3275-5 (pbk.)
        1. System analysis. 2. Decision making. 3. Information technology. I. Popper, Steven W., 1953– II. Bankes, Steven C. III.Title.

T57.6 .L46 2003
320'.6'0113—dc21

                                                                    2003012438

Cover design by Barbara Angell Caslon

RAND is a nonprofit institution that helps improve policy and decisionmaking through research and analysis. RAND® is a registered trademark. RAND's publications do not necessarily reflect the opinions or policies of its research sponsors.

Published 2003 by RAND
1700 Main Street, P.O. Box 2138, Santa Monica, CA 90407-2138
1200 South Hayes Street, Arlington, VA 22202-5050
201 North Craig Street, Suite 202, Pittsburgh, PA 15213-1516
RAND URL: http://www.rand.org/
To order RAND documents or to obtain additional information, contact Distribution Services: Telephone: (310) 451-7002;
Fax: (310) 451-6915; Email: order@rand.org

Is there any practical value in considering the long term—25, 50, or 100 years into the future—when debating policy choices today? If so, how is it possible to use these considerations to actually inform the actions we will take in the near term? This study is an initial effort by the RAND Pardee Center to frame a role for long-term policy analysis. It considers the history of attempts to treat the future in an analytical manner and then offers a new methodology, based on recent advances in computer science, that shows promise for making such inquiries both practicable and useful. It suggests a new approach for systematic consideration of a multiplicity of plausible futures in a way that will enhance our ability to make good decisions today in the face of deep uncertainty.

This research was undertaken through a generous gift from Frederick S. Pardee to develop improved means of systematically dealing with the uncertainties of a longer-range future. This report should be of interest to decisionmakers concerned with the long-term effects of their actions, those who conduct long-term planning, and anyone who deals more generally with decisionmaking under deep uncertainty. The report should also interest those concerned with the latest advances in computer technology in support of human reasoning and understanding.

ABOUT THE RAND PARDEE CENTER

The RAND Frederick S. Pardee Center for Longer Range Global Policy and the Future Human Condition was established in 2001 through a gift from Frederick S. Pardee. The Pardee Center seeks to enhance the overall future quality and condition of human life by improving longer-range global policy and long-term policy analysis. In carrying out this mission, the center concentrates on five broad areas:

- Developing new methodologies, or refining existing ones, to improve thinking about the long-range effects of policy options.

- Developing improved measures of human progress on a global scale.

- Identifying policy issues with important implications for the long-term future—i.e., 35 to 200 years ahead.

- Using longer-range policy analysis and measures of global progress to improve near-term decisions that have long-term impact.

- Collaborating with like-minded institutions and colleagues, including international organizations, academic research centers, futures societies, and individuals around the globe.

Inquiries regarding the RAND Pardee Center may be directed to

James A. Dewar
Director
RAND Pardee Center
1700 Main Street
Santa Monica, CA 90401
Phone: (310) 393-0411 extension 7554
E-mail: dewar@rand.org
Web site: http://www.rand.org/pardee/

CONTENTS

FIGURES

TABLES

New analytic methods enabled by the capabilities of modern computers may radically transform human ability to reason systematically about the long-term future. This opportunity may be fortuitous because our world confronts rapid and potentially profound transitions driven by social, economic, environmental, and technological change. Intentionally or not, actions taken today will influence global economic development, the world's trading system, environmental protection, the spread of such epidemics as AIDS, the fight against terrorism, and the handling of new biological and genetic technologies. These actions may have far-reaching effects on whether the twenty-first century offers peace and prosperity or crisis and collapse.

In many areas of human endeavor, it would be derelict to make important decisions without a systematic analysis of available options. Powerful analytic tools now exist to help assess risks and improve decisionmaking in business, government, and private life. But almost universally, systematic quantitative analysis rarely extends more than a few decades into the future. Analysts and decisionmakers are neither ignorant of nor indifferent to the importance of considering the long term. However, well-publicized failures of prediction—from the Club of Rome's "Limits to Growth" study to the unexpected, sudden, and peaceful end of the Cold War—have done much to discourage this pursuit. Systematic assessments of the long-term future are rare because few people believe that they can be conducted credibly.

A PROSTHESIS FOR THE IMAGINATION

This report describes and demonstrates a new, quantitative approach to long-term policy analysis (LTPA). These robust decisionmaking methods aim to greatly enhance and support humans' innate decisionmaking capabilities with powerful quantitative analytic tools similar to those that have demonstrated unparalleled effectiveness when applied to more circumscribed decision problems. By reframing the question "What will the long-term future bring?" as "How can we choose actions today that will be consistent with our long-term interests?" robust decisionmaking can harness the heretofore unavailable capabilities of modern computers to grapple directly with the inherent difficulty of accurate long-term prediction that has bedeviled previous approaches to LTPA.

This report views long-term policy analysis as a way to help policymakers whose actions may have significant implications decades into the future make systematic, well-informed decisions. In the past, such decisionmakers, using experience, a variety of heuristics, rules of thumb, and perhaps some luck, have occasionally met with impressive success, for example, in establishing the West's Cold War containment strategy or in promoting the first U.S. transcontinental railroads to forge a continent-sized industrial economy. Providing analytic support to improve such decisionmaking must contend with a key defining feature of the long term—that it will unavoidably and significantly be influenced by decisions made by people who live in that future. **Thus, this study defines the aim of LTPA as identifying, assessing and choosing among near-term actions that shape options available to future generations.**

LTPA is an important example of a class of problems requiring decisionmaking under conditions of _deep uncertainty_—that is, where analysts do not know, or the parties to a decision cannot agree on, (1) the appropriate conceptual models that describe the relationships among the key driving forces that will shape the long-term future, (2) the probability distributions used to represent uncertainty about key variables and parameters in the mathematical representations of these conceptual models, and/or (3) how to value the desirability of alternative outcomes. In particular, the long-term future may be dominated by factors that are very different from the current drivers and hard to imagine based on today's experiences. Meaningful LTPA must confront this potential for surprise.

Advances in LTPA rest on solid foundations. Over the centuries, humans have used many means to consider both the long-term future and how their actions might affect it. Narratives about the future, whether fictional or historical, are unmatched in their ability to help humans viscerally imagine a future different from the present. Such group methods as Delphi and Foresight exploit the valuable information often best gathered through discussions among groups of individuals. Analytic methods—e.g., simulation models and formal decision analyses—help correct the numerous fallacies to which human reasoning is prone. Scenario planning provides a framework for *what if–ing* that stresses the importance of multiple views of the future in exchanging information about uncertainty among parties to a decision. Despite this rich legacy, all these traditional methods founder on the same shoals. The long-term future presents a vast multiplicity of plausible futures. Any one or small number of stories about the future is bound to be wrong. Any policy carefully optimized to address a "best guess" forecast or well-understood risks may fail in the face of inevitable surprise.

This study proposes four key elements of successful LTPA:

- Consider large *ensembles* (hundreds to millions) of scenarios.

- Seek *robust,* not optimal, strategies.

- Achieve robustness with *adaptivity.*

- Design analysis for *interactive exploration* of the multiplicity of plausible futures.

These elements are implemented through an iterative process in which the computer helps humans create a large ensemble of plausible scenarios, where each scenario represents one guess about how the world works (a future state of the world) and one choice of many alternative strategies that might be adopted to influence outcomes. Ideally, such ensembles will contain a sufficiently wide range of plausible futures that one will match whatever future, surprising or not, does occur—at least close enough for the purposes of crafting policies robust against it. Robust decisionmaking then exploits the interplay between interactive, computer-generated visualizations called "landscapes of plausible futures" that help humans form hypotheses about appropriate strategies and computer searches across the ensemble that systematically test these hypothesis.

In particular, rather than seeking strategies that are optimal for some set of expectations about the long-term future, this approach seeks near-term strategies that are robust—i.e., that perform reasonably well compared to the alternatives across a wide range of plausible scenarios evaluated using the many value systems held by different parties to the decision. In practice, robust strategies are often adaptive; that is, they evolve over time in response to new information. Adaptivity is central to the notion that, when policymakers consider the long term, they seek to shape the options available to future generations. Robustness reflects both the normative choice and the criterion many decisionmakers actually use under conditions of deep uncertainty. In addition, the robustness criterion is admirably suited to the computer-assisted discovery and testing of policy arguments that will prove valid over a multiplicity of plausible futures.

At its root, robust decisionmaking combines the best capabilities of humans and computers to address decision problems under conditions of deep uncertainty. Humans have unparalleled ability to recognize potential patterns, draw inferences, formulate new hypotheses, and intuit potential solutions to seemingly intractable problems. Humans also possess various sources of knowledge—tacit, qualitative, experiential, and pragmatic—that are not easily represented in traditional quantitative formalisms. Humans also excel, however, at neglecting inconvenient facts and at convincing themselves to accept arguments that are demonstrably false. In contrast, computers excel at handling large amounts of quantitative data. They can project without error or bias the implications of those assumptions no matter how long or complex the causal chains, and they can search without prejudice for counterexamples to cherished hypotheses. Working interactively with computers, humans can discover and test hypotheses about the most robust strategies. Thus, computer-guided exploration of scenario and decision spaces can provide a prosthesis for the imagination, helping humans, working individually or in groups, to discover adaptive near-term strategies that are robust over large ensembles of plausible futures.

DEMONSTRATING ROBUST DECISIONMAKING   *Example*

This study demonstrates new robust decision methods on an archetypal problem in long-term policy analysis—that of global sustainable development. This topic is likely to be crucially important in the twenty-first century. It is fraught with deep uncertainty. It incorporates an almost unmanageably wide range of issues, and it engages an equally wide range of stakeholders with diverse values and beliefs. This sustainable-development example demonstrates the potential of robust decisionmaking to help humans reason systematically about the long-term implications of near-term actions, to exploit available information efficiently, and to craft potentially implementable policy options that take into account the values and beliefs of a wide variety of stakeholders.

The project team began by reviewing and organizing the relevant background information, particularly from the extensive literature on sustainability. The team also assembled a group of RAND experts to act as surrogate stakeholders representing a range of opinions in the sustainability debate. To help guide the process of elicitation and discovery and to serve as an intellectual bookkeeping mechanism, the study employed an "XLRM" framework often used in this type of analysis The key terms are defined below.[1]

- Policy levers ("L") are near-term actions that, in various combinations, comprise the alternative strategies decisionmakers want to explore.

- Exogenous uncertainties ("X") are factors outside the control of decisionmakers that may nonetheless prove important in determining the success of their strategies.

- Measures ("M") are the performance standards that decisionmakers and other interested communities would use to rank the desirability of various scenarios.

- Relationships ("R") are potential ways in which the future, and in particular those attributes addressed by the measures, evolve

---

[1]This discussion continues the long-standing practice of ordering the letters XLRM. However, in this instance, a clearer exposition was achieved by presenting the factors in a different order.

over time based on the decisionmakers' choices of levers and the manifestation of the uncertainties. A particular choice of Rs and Xs represents a future state of the world.

In the approach described in this report, the first three factors—near-term actions (L), uncertainties (X), and performance measures (M)—are tied together by the fourth (R), which represents the possible relationships among them. This decision-support system thus becomes a tool for producing interactive visual displays (i.e., landscapes of plausible futures) of the high-dimensional decision spaces inherent in LTPA problems. The system employs two distinct types of software:

- *Exploratory modeling software* enables users to navigate through the large numbers of scenarios required to make up a scenario ensemble and to formulate rigorous arguments about policy choices based on these explorations.

- *A scenario generator* uses the relationship among the variables to create members of scenario ensembles. In contrast to a traditional model that is typically designed to produce a comparatively small number of predictive conclusions, a scenario generator should yield a full range of plausible alternatives .

In combination, these two types of software enable humans to work interactively with computers to discover and test hypotheses about robust strategies.

The robust decision analysis reported in this study begins with a diverse scenario ensemble based on XLRM information. A modified version of the "Wonderland" system dynamics model functions as the scenario generator. The analysis examines and rejects a series of candidate robust strategies and, by appropriate use of near-term adaptivity, it eventually arrives at a promising near-term policy option. The robust strategy sets near-term (10-year) milestones for environmental performance and adjusts policies annually to reach such milestones, contingent on cost constraints. Compared to the alternatives, it performs well over a wide range of plausible futures, using four different value systems for ranking desirable futures.

A steering group of surrogate stakeholders was then challenged to imagine surprises representing distinct breaks with current trends or expectations. These surprises were added to the scenario generator and the policy options stress-tested against them. The analysis concludes by characterizing the wager decisionmakers would make if they choose not to hedge against those few futures for which the proposed robust strategy is not an adequate response. This iterative process thus provides a template for designing and testing robust strategies and characterizing the remaining "imponderable" uncertainties to which they may be vulnerable.

Not enough detail of example to be convincing !

Residual Risk

SEIZING THE NEW OPPORTUNITIES FOR LTPA

This report does not provide specific policy recommendations for the challenge of sustainable development. The analysis involves neither the level of detail nor the level of stakeholder participation necessary for policy results that can be acted on. Rather, the study aims to describe the new analytic capabilities that have become available to support long-term decisionmaking. The report concludes with a description of how future work might improve on the robust decision approach to LTPA as well as some of the challenges and potential suggested by this limited demonstration. In particular, policy-relevant LTPA will require improved scenario generators, better algorithms to support navigation through large scenario ensembles, improved treatment of measures of the future human condition, and refined protocols for engaging the parties in a decision in a robust policymaking exercise and widely disseminating the results.

Challenges

The lack of systematic, quantitative tools to assess how today's actions affect the long-term future represents a significant missed opportunity. It creates a social context where values relating to long-term consequences cannot be voiced easily because they cannot be connected to any practical action. Across society, near-term results are often emphasized at the expense of long-term goals. However, our greatest potential influence for shaping the future may often be precisely over those time scales where our gaze is most dim. By its nature, where the short term is predictable and subject to forces we can quantify, we may have little effect. Where the future is ill-defined, hardest to see, and pregnant with possibilities, our actions may well have their largest influence in shaping it.

Only in the last few years have computers acquired the power to support directly the patterns of thought and reason humans traditionally and successfully use to create strategies in the face of unpredictable, deeply uncertain futures. In today's era of radical and rapid change, immense possibilities, and great dangers, it is time to harness these new capabilities to help shape the long-term future.

ACKNOWLEDGMENTS

We have spent much of the last decade struggling with the related questions of how to craft methods for decisionmaking under deep uncertainty and finding the value of computer simulations in situations where it is obvious any predictions will be wrong. Along the way we have drawn inspiration and good advice from many colleagues at RAND and elsewhere, including Carl Builder, Thomas Schelling, James Hodges, John Adams, David Robalino, and Michael Schlesinger. One of the great pleasures of this particular project has been the much-welcomed opportunity to work closely with James Dewar. His seminal work on Assumption Based Planning provides one key inspiration for our work with robust decision methods, and his input during this project has been that of a thoughtful, encouraging, and engaged colleague.

Frederick Pardee's passion for improving the long-term future human condition provided the support for this work. Fred has made an important and astute choice for his philanthropy. He understands that the overwhelming focus of government, business, and most foundations on the short term may blind society to some of the most important and much-needed actions we could take today to shape the decades ahead. We hope that use of the robust decision methods we describe in this study may make systematic and effective thinking about the long-term future far more common and enable many to blaze the path that Fred has envisioned.

Many colleagues have contributed to the work described here. RAND graduate fellow Kateryna Fonkych helped with explorations of the International Futures and Wonderland scenario generators, fellow David Groves assisted with data analysis, and fellow Joseph Hendrickson helped with the analytic methods for navigating through scenario spaces discussed in Chapter Six. Our advisory group—Robert Anderson, Sandra Berry, Robert Klitgaard, Eric Larson, Julia Lowell, Kevin McCarthy, David Ronfeldt, and George Vernez—gave generously of their time and provided numerous inputs of valuable advice. Our reviewers, William Butz, Al Hammond, and Bruce Murray, offered well-targeted suggestions that did much to improve our manuscript. Caroline Wagner offered many probing questions as we initially formulated this effort.

Judy Larson proved invaluable in shepherding three authors with different styles toward a unified prose and in gathering 10 years of musings into a single story. Our editor, Dan Sheehan, helped turn Word files into a published document, and Mary Wrazen helped mold computer printouts into presentable graphics.

Additional funding for the analytic methodology development was provided by the U.S. National Science Foundation under Grant BCS-9980337 and the Defense Advanced Projects Research Agency. Evolving Logic provided the CARs™ software used to support this project.

We hope that this work helps many others launch their own explorations into how today's actions can best shape our long-term future. We accept full and sole responsibility for any errors remaining in this report.

ABBREVIATIONS

CARs™	Computer-Assisted Reasoning® system by Evolving Logic
CPU	Central Processing Unit
GDP	Gross domestic product
GSG	Global Scenarios Group
HDI	Human Development Index
ICIS	International Centre for Integrative Studies
ICSU	International Council of Scientific Unions
IFs	International Futures computer simulation by Barry Hughes
IPCC	Intergovernmental Panel on Climate Change
LTPA	Long-term policy analysis
NISTEP	National Institute of Science and Technology Policy
NRC	Nuclear Regulatory Commission
OECD	Organization for Economic Cooperation and Development
PPP	Purchasing power parity
RAP™	Robust Adaptive Planning by Evolving Logic
SRES	Special Report on Emissions Scenarios
UNDP	United Nations Development Programme
XLRM	A framework that uses exogenous uncertainties, policy levers, relationships, and measures

Chapter One

THE CHALLENGE OF LONG-TERM POLICY ANALYSIS

Our world confronts rapid and potentially profound transitions driven by social, economic, environmental, and technological change. Countries that have achieved political stability and wealth coexist uneasily among regions with fragile governments and economies whose people often live in dire poverty. Pressures grow on the natural environment. Technology has created tremendous opportunities but has also unleashed awesome destructive power more readily accessible than imagined a few decades ago. It is increasingly clear that today's decisions could play a decisive role in determining whether the twenty-first century offers peace and prosperity or crisis and collapse.

In many areas of human endeavor one would be derelict in making important decisions without undertaking a systematic analysis of the available options. Before investing in a new business venture, managing a large financial portfolio, producing a new automobile, deploying a modern army, or crafting a nation's economic policy one would identify a range of alternatives and use available information to make quantitative comparisons of the likely consequences of each alternative.

However, beyond a certain time horizon quantitative analysis is rarely attempted. For example, quantitative modeling of national economic performance informs fiscal policy only a few quarters away. In business planning, time frames longer than one year are considered strategic. Military planning looks farther ahead, yet defense analysis directed more than 10 years into the future is rare and longer than 15 years is virtually nonexistent. Civic planning

1

sometimes, but not often, encompasses two decades. Official government forecasts of energy production and consumption rarely extend beyond 20 years.

This is not to say that analysts and decisionmakers are ignorant of or indifferent to the importance of planning for the long term. In some cases, people have taken actions intended to shape the long-term future and have on occasion met with impressive success. At the start of the Cold War, for example, the United States and its allies laid out a plan to defeat Soviet Communism by containing its expansion until the system ultimately collapsed from its own internal contradictions (Kennan, 1947). This policy was often implemented in forms that differed from the original design, was on occasion invidious to some developing countries' aspirations for self-determination, and produced moments when the world was closer to nuclear war than anyone could wish. Nonetheless, through a combination of good planning, skillful implementation, and luck, the policy worked after 40 years almost exactly as intended. Similarly, U.S. policymakers in the late 1860s offered massive financial incentives for entrepreneurs to build risky and expensive rail lines across North America (Bain, 1999). While this policy launched a process rife with amazing determination, thievery, heroism, cruelty, and corruption, over the following decades it accomplished precisely what was intended. The transcontinental railroad stitched together a nation recently shattered by civil war and enabled the world's first, and still the strongest, continental industrial economy.

Of course, in many cases decisionmakers deem potential long-term benefits less important than such immediate concerns as the results of the next election or an upcoming quarterly report to shareholders. But even when decisionmakers obviously value the long term, they are often uncertain about how to translate their concerns into useful action. Broadly speaking, people do not conduct systematic, long-term policy analysis (LTPA) because no one knows how to do it credibly.

The inability of the policy and analytic communities to plan for the long term in a manner perceived as rigorous, credible, and demonstrably useful has major consequences for society. The lengthy history of failed forecasts encourages a general belief that it is pointless to think about a far future that cannot be predicted with any degree of assurance. This creates a social context in which values relating to long-term consequences cannot be voiced easily because they cannot be connected to any practical action. Thus, there is a general tendency across the social spectrum to emphasize near-term results at the expense of long-term goals. Paradoxically, people often have a great deal of analytic support for short-term decisions, many of which may be easily adjusted when new information suggests a need to change course. When they make decisions with long-term consequences, potentially shaping the world they and their descendants will occupy for decades, people are, in effect, flying blind.

QUANTITATIVE LTPA MAY NOW BE POSSIBLE

For the purposes of this report, long-term policymakers are those who consider the implications of their actions stretching out many decades into the future. Stated another way, **long-term policymaking takes place when the menu of near-term policy options considered by decisionmakers and the choices they make from that menu are significantly affected by events that may occur 30 or more years into the future.**

LTPA helps policymakers make systematic, well-informed, long-term policy decisions. As discussed in later chapters, a key defining feature of the long term is that it will be influenced unavoidably and significantly by decisions made by people in the future. Thus, **LTPA aims to identify, assess, and choose among near-term actions that shape options available to future generations.**

There are many types of LTPA. In this report, we focus on quantitative methods similar to those that have proved so indispensable for other types of decision problems—that is, ones that rely on data and known laws of logical, physical, and social behavior expressed in mathematical form.

Deep Uncertainty Challenges LTPA

LTPA is an important example of a class of problems requiring decisionmaking under conditions of deep uncertainty. Deep uncertainty exists when analysts do not know, or the parties to a decision cannot agree on, (1) the appropriate models to describe the interactions among a system's variables, (2) the probability distributions to represent uncertainty about key variables and parameters in the models, and/or (3) how to value the desirability of alternative outcomes.[1]

Humans often confront conditions of deep uncertainty. They frequently respond successfully, provided that their intuition about the system in question works reasonably well. Often, decisionmakers identify patterns based on a wealth of past experience that suggest an appropriate response to some new situation. For instance, seasoned decisionmakers, such as fire chiefs arriving at the scene of a blaze, will rapidly classify a situation as some familiar type: Is this a case where people may be trapped inside a building, where the building may collapse, where the fire can be extinguished, or where it can merely be contained? Next, they choose an appropriate course of action and run a mental simulation to test their plan against the particulars of the situation before them (Klein, 1998).[2] Humans may also employ heuristics, or rules of thumb, to serve as quick surrogates for complex calculations. Many firms will adjust the hurdle rate for the return on investment required to go forward with a large capital project in response to changes in market opportunities or the state of the economy (Lempert et al., 2002). Humans have also developed iterative, sometimes collaborative processes to produce and test plans under such conditions and they likewise have the procedures and institutions for implementing them. Capitalizing on a facility for storytelling, U.S. officials during the Cuban Missile Crisis debated alternative courses of action by challenging each other with

---

[1] A number of different terms are used for concepts similar to what we define as deep uncertainty. Knight (1921) contrasted risk and uncertainty, using the latter to denote unknown factors poorly described by quantifiable probabilities. Ellsberg's (1961) paradox addresses conditions of ambiguity where the axioms of standard probabilistic decision theory need not hold. There is an increasing literature on ambiguous and imprecise probabilities (de Cooman, Fine, and Seidenfeld, 2001). Ben-Haim's (2001) Info-Gap approach addresses conditions of what he calls severe uncertainty. We take the phrase "'deep'" uncertainty from a presentation by Professor Kenneth Arrow (2001) describing the situation faced by climate change policymakers. The precise definition of this term is our own.

[2] This report's definition of long-term policymaking assumes that decisionmakers are operating in a mode in which they lay out several options, assess their consequences, and choose among them—that is, that they choose differently from the fire chief described here. Nonetheless, the quantitative methods proposed in this study build on this human ability to draw inferences from recognized patterns and test strategies with mental simulations.

"what if" scenarios to probe for weaknesses in proposed plans (Allison and Zelikow, 1999).[3] These processes frequently succeed because the best response to deep uncertainty is often a strategy that, rather than being optimized for a particular predicted future, is both well-hedged against a variety of different futures and is capable of evolving over time as new information becomes available.

The process of mining experiential information and repeatedly examining proposed strategies over a range of contingencies can, however, easily break down, especially when humans are confronting novel conditions or extensive amounts of information. In such situations, humans rapidly lose the ability to track long causal links or the competing forces that may drive the future along one path or another. Biases may focus undue attention on expected futures or the performance of desired strategies. The human ability to recognize the correct patterns or trace the "what if" implications of proposed plans may quickly prove inadequate to such challenges.

The quantitative tools of decision analysis can help people systematically assess the implications of extensive information and expose biases and flaws in their reasoning.[4] Under conditions of deep uncertainty, however, the application of these traditional quantitative methods is fraught with problems. At the most basic level, the process may simply terminate in gridlock if more than one individual is responsible for making the decision and the participants cannot agree on the assumptions that will form the basis of the analysis. Even if this hurdle is overcome and candidate strategies are forth-

---

[3]The main theme of Allison's famous book, first published in 1972, that "multiple, overlapping, competing conceptual models are the best that the current understanding of foreign policy provides" (p. 401) resonates with the type of uncertainty this study report aims to address.

[4]During the past 50 years, statisticians and operations researchers have developed a host of powerful analytic techniques for addressing uncertainty and risk management. The tools have a wide variety of names, but fundamentally they are based on the concepts of Bayesian decision analysis. This approach assumes that knowledge about the future may be described with a system model that relates current actions to future outcomes and that uncertainty may be described by subjective probability distributions over key input parameters to the model. (For a review of these methods see Morgan and Henrion, 1990). Many of these tools were originally developed in the 1950s when computer power was meager. Thus, to reduce the computational burden, they placed a premium on reducing information about the future into a small set of best estimates.

coming, a traditional approach is likely to suggest policies that may prove brittle against surprise or unworkable in application. Most policymakers recognize that a deeply uncertain long-term future is sure to offer surprises. Policies put forth by traditional quantitative methods may perform poorly in the face of unexpected contingencies and thus provide a poor guide to shaping the long term.

Modern Computational Power Creates New Possibilities

When human intuition about cause and effect breaks down, mathematics and computers can become crucial supports to decisionmaking. This report argues that new capabilities conferred by modern computers may now enable useful and relevant LTPA. The widespread availability of fast processing, virtually unlimited memory, and interactive visualizations can link the innate human capacity for heuristics with powerful quantitative analytic tools that have demonstrated unparalleled effectiveness in dealing with more circumscribed decision problems.

Traditional quantitative tools use the computer as a calculator. Humans assemble data and assumptions and feed them into the computer, which then reports what appears to be the most desirable strategy based on the limited data provided. This approach encourages people to narrow the range of their speculations so that the analysis can recommend a *definitive* course of action.

In contrast, new robust decision methods use the computer as an interactive tool to help people think creatively across the multiplicity of futures they face and come to concrete conclusions about the best ways of shaping those futures to their liking. The computer can then be used to test those conclusions systematically against the full range of available information.

Under conditions of deep uncertainty, we suggest that analysts use computer simulations to generate a large ensemble of plausible scenarios about the future.[5] Each scenario represents one guess about

---

[5]The methods described in this report can also be used with statistical models, neuralnets, and other mathematical representations that unlike simulation models do not contain explicit assumptions about causality by using such mathematical representations to create multiple fits to available data.

how the world works and one choice among many alternative strategies that might be adopted to influence outcomes. In an integrated division of labor, the computer generates visualizations that allow humans to form hypotheses about their best decisions. As part of the reasoning process, the computer is then used to conduct searches systematically across the scenarios to test these hypotheses. The goal is to discover near-term policy options that are robust over a wide range of futures when assessed with a wide range of values. Robust strategies will often be adaptive—that is, they will be explicitly designed to evolve over time in response to new information.[6]

THE CHALLENGE OF GLOBAL SUSTAINABLE DEVELOPMENT

The basic approach described in this report has been applied to problems in defense, government, and business. Here, we present the first complete application to LTPA. In the course of our discussion, we will address the typical *why* and *how* questions that emerge when LTPA and strategic decisionmaking intersect: "Why bother looking at the long-term future when accurate prediction is not possible?" "How can considerations of the long-term future be credibly incorporated into serious deliberations about policy?"

For purposes of demonstration, this report centers on the issue of global sustainable development, a paradigmatic candidate for LTPA. This topic is likely to be crucially important in the twenty-first century. It is fraught with deep uncertainty. It incorporates an almost unmanageably wide range of issues, and it engages an equally wide range of stakeholders with diverse values and beliefs. We do not claim to have solved the problem. Rather, through this example we

---

[6] The robust decisionmaking approach is related to the Monte Carlo analyses increasingly applied to decisionmaking and risk assessment. As generally employed, Monte Carlo analysis scans over a large number of plausible futures by assuming probability distributions for the uncertainties in key input parameters to some system model. The computer then randomly samples some of these inputs and calculates a probability distribution of outputs. These output distributions may be used to calculate the expected value of alternative policy options and/or the risk (that is, likelihood) of various adverse outcomes. In contrast, robust decision approaches use the computer to scan over many plausible futures to identify those that may be particularly useful to humans in designing and stress testing robust strategies. Monte Carlo sampling is one type of method that can be used to identify such futures.

intend to show that it is possible to reason about the long-term implications of near-term actions, to exploit available information efficiently, and to craft potentially implementable policy options that take into account the values and beliefs of a wide variety of stakeholders.

SURPRISE: THE CONSTANT ELEMENT

One assertion about a deeply uncertain long-term future would seem to be inarguable. No matter how inclusive the information-gathering efforts, how effective the analytic tools and techniques, how profound our insights, and how careful the resulting preparations, the future is certain to follow paths and offer events we did not imagine. Surprise takes many forms, all of which tend to disrupt plans and planning systems.

However, this very certainty of surprise underscores the advantages of the robust decision method for conducting LTPA. Rather than offering predictions about the future, an iterative, interactive approach provides the analytic framework for encouraging people and groups to think systematically and creatively about potential surprises and the possible responses to them (Lempert, Popper, and Bankes, 2002). The approach employs a diverse collection of plausible futures to stress test candidate strategies and to help discover policy options demonstrably robust to known uncertainties. It is through robustness, whether obtained from adaptability or armoring, that biological organisms and human institutions can survive surprises. Although it will never be possible to anticipate every surprise before it happens, the method described here can greatly increase the likelihood that policymakers have chosen actions that are robust against whatever the future has in store.

ORGANIZATION OF THIS REPORT

This report is intended for decisionmakers who may wish to improve their ability to shape the long term, policy analysts who wish to assess a new approach they might wish to add to their toolkit, and the lay reader interested in new ways to understand and influence the future we shall all inhabit. As with any such document that addresses multiple audiences, different readers will find different parts of greater interest.

Chapter Two briefly surveys the principal means humans have traditionally used to ponder the problem of long-term decisionmaking. It lays out the common, main stumbling block—an inability to address a multiplicity of plausible futures. Chapter Three presents a new robust decisionmaking approach to LTPA. This approach combines modern computer technology with the innate capacities of the human mind in an iterative process that discovers and repeatedly tests near-term strategies robust against a large ensemble of plausible futures. The information in both of these chapters will, we hope, prove useful to all readers.

Chapters Four and Five describe in detail a demonstration application of LTPA to the problem of global sustainable development in the twenty-first century. This demonstration employed only a simple set of models and data and engaged only a small group of surrogates representing larger stakeholder groups. Chapter Six suggests appropriate next steps for expanding this demonstration to produce policy-relevant results. More technical in nature, these chapters should prove most relevant to analysts and analytically inclined decisionmakers whose responsibilities require them, on the one hand, to gather and interpret data and, on the other, to make decisions that have implications for the long term.

Chapter Seven offers some summary observations that should be accessible and helpful to all readers.

The appendices describe the "Wonderland" scenario generator used in this study, and they also supply supporting detail for the analysis presented in Chapter Five. This nuts-and-bolts material should primarily interest members of the modeling and simulation communities and analysts who seek deeper insight into the new approach to LTPA described in this report.

Chapter Two

A HISTORY OF THINKING ABOUT THE FUTURE

Interest in the future is not new. Human reason and imagination have always compelled people to reflect on the past and speculate on what will be. This chapter surveys the principal means humans have used over the millennia to consider the long-term future and how their actions might affect it. This broad view and a focus on the essence of each approach leads to two basic findings. The first provides a source of comfort. Tools that support thinking about the long-term consequences of today's actions have a lengthy pedigree. Much has been done, providing a trove of experience and insight from which to draw. This rich heritage enables consideration of meaningful LTPA and provides the foundation for the rest of the discussion to follow.

At the same time, a second theme suggests the key challenge. Despite the often profound capabilities any traditional method provides, none supports a truly satisfactory LTPA. All suffer a common weakness—the inability to come to grips with the multiplicity of alternative plausible futures. Clearly, LTPA must struggle with this central problem no matter what the actual substance of the analysis.

This chapter will briefly highlight the many strengths and this central weakness of the traditional methods for LTPA. The rest of the report will argue that modern computer technology can break through previous constraints. In particular, the unprecedented capability of modern computers to handle a huge ensemble of plausible futures offers a means to exploit the profound insights from the traditional methods for thinking about the future and weave them into a powerful new approach to LTPA.

NARRATIVES: MIRRORS OF THE PRESENT, VISIONS OF THE FUTURE

Narratives about the future are an extraordinarily powerful means of engaging the imagination. From earliest times, storytelling[1] was the principal vehicle for developing and communicating explanations of the way things were and how they came to be. It was also a tool for addressing anxiety about matters related to future survival—that is, if one could somehow acquire information about events that were likely to occur, it might be possible to prepare for them and to achieve desirable outcomes.

For many centuries, seers and prophets have provided descriptions of the future to help human beings understand their place in the universe and to suggest codes of behavior and courses of action consistent with that knowledge. At the highest levels of policy, this is also a course prudence suggested. Such narratives often took the forms of oracles. King Saul consulted the Witch of Endor against the specific proscriptions of the prophets of Israel; that he did so indicates both the power of belief and the anxiety he felt about future outcomes. When the elders of Pericles's Athens received (typically cryptic) forewarning of the coming Peloponnesian War from the oracle at Delphi, they were engaging in what the norms of their time held to be due diligence.

Formal fictional forms have considered the future. Written accounts of utopias—ideal societies whose citizens live in a condition of harmony and well-being—date back at least as far as Plato's *Republic* (c. 360 BC). Perhaps the best-known American example of a utopian work is Edward Bellamy's *Looking Backward, 2000–1887* (published in 1888), where a nineteenth century man awakes to find himself transported to Boston in the year 2000. There he encounters a socialistic society in which inequities of education, health care,

---

[1] The process of mythmaking is relevant in this context. As noted in the *Encyclopedia Britannica*, myth "has existed in every society... [and] would seem to be a basic constituent of human culture." Unburdened by a requirement for empirical proof, myths offer comprehensive explanations of the natural and supernatural worlds and mankind's relationship to both. A point of particular interest for readers of this report is the assertion that "The function of models in physics, biology, medicine, and other sciences resembles that of myths as paradigms, or patterns, of the human world" (http://www.search.eb.com/eb/article?eu=115608>).

models function as myths! Interesting viewpoint

career opportunity, social status, and material wealth have been engineered out of existence.[2] In more recent times, science fiction has used the dynamics of social and scientific-technical change as a springboard to explore the currents propelling people away from their familiar worlds.[3]

From the perspective of LTPA, the principal value of narratives is that they provide a tool to help people confront the long-term future and frame what appear reasonable courses of action by imagining what it may be like to live there. It is exactly the relationship between near-term actions and long-term consequences that is the crux of LTPA. A classic modern example is Rachel Carson's *Silent Spring* (1962), a vivid depiction of a future world whose wildlife has been exterminated by pollution. *Silent Spring* was a best seller, and it had the desired effect of helping to spark a worldwide movement in support of societal action for environmental protection. Yet, like *Silent Spring*, most futuristic narratives are created with the aim of commenting on and shaping the present rather than supplying an accurate roadmap for what is to come.

Lessons from History Can Help Anchor Speculations About the Future

The obvious problem with using narratives about the long term to inform present-day actions is that while these stories may offer compelling, insightful commentary about current options they are usually wrong in many important details about the future. Cognizant of this deficiency, people who wish to develop and communicate their ideas about the future have tried several techniques to improve the narrative approach to LTPA. Relying on the lessons of history provides one means of grounding narrative predictions. Because, in the broadest sense, history is the story of the past, it contains a mother lode of data relevant to what may be. It also offers a temporal vantage point that sets some bounds on the extent to which things may change or stay the same over decades and centuries.

---

[2]Of course, not all visions of the future were so blissful nor were planned societies so appealing. Aldous Huxley's *Brave New World* (1932) and George Orwell's *1984* (1949) are powerful examples of "dystopias."

[3]For more discussion see, for example, Aldiss (1986) and Alkon (1987).

In applying knowledge of history, some analysts focus on a specific period in the past and draw parallels to contemporary and future times. For instance, James A. Dewar (1998) attempted to understand the potential social consequences of the Internet by examining the social effects of the printing press. Among the most significant of those consequences was the printing press's dramatic reduction of the cost and scope of one-to-many communication. This leap in technological capability, Dewar argued, led to profound changes in human society ranging from the Reformation to the scientific revolution. He then observed that the Internet for the first time allows many-to-many communication on a global scale, and he asserted that this capability is of similar magnitude to that of the printing press. Rather than formulating specific predictions, Dewar used historical parallels as the basis for inferences regarding forces that could bear importantly on the information revolution. Such insights may be used to suggest points to consider in framing long-term policy.[4]

Attempting to discern key historical trends is another way of apprehending the long-term future. This approach to history ideally leads to detection, then interpretation, of large-scale patterns or "grand designs," which become the basis for prediction by extension. The ancient Chinese, Hindus, Greeks, and Mayans all noted archetypal patterns in time. To them, history represented a series or recurrence of alternating phases where periods of unity and peace were succeeded by division and disintegration, followed by rehabilitation and restoration of harmony, perhaps occurring on a higher plane than before. Later philosopher-historians pursued a similar concept. In early eighteenth century Italy, Giambattista Vico described the successive stages of growth and decay that characterize human societies. G. W. F. Hegel developed his *dialectical* concept of thesis, antithesis, and synthesis. Nineteenth century thinkers Karl Marx and Friedrich Engels placed Hegel's philosophy in a more distinctly social

---

[4]Dewar (1998) stated that the main social ramifications of the printing press were unintended, and he believed that the information revolution would be similarly dominated by unpredictable and unintended consequences. Therefore, he posited two general lessons for information-age policymakers. First, noting that those countries benefiting most from the printing press regulated it least, Dewar argued that the Internet should remain unregulated. Second, he suggested that policy toward the Internet should emphasize experimentation as well as quick exposure of and response to unintended consequences.

context through their elaboration of the continual struggle between the proletariat and the bourgeoisie that would one day culminate in a classless society, the overthrow of capitalism, and the elimination of organized government. In the twentieth century, Oswald Spengler contended that human civilizations followed the path of natural organisms in a pattern of birth, development, and decay. In contrast, Arnold Toynbee believed that civilizations grew and prospered by responding to a series of challenges, and he did not share Spengler's notion that rejuvenation was impossible. The retrospective inability of these grand architectures of history to anticipate the changes the world has already witnessed has caused interest in this approach to speculating about the future to decline in recent years.

Herman Kahn's treatise, _The Next 200 Years_ (1976), is a more contemporary example of this approach. In a sweeping narrative, Kahn sought to ground his speculations in careful quantitative analysis of historical data and potential future trends.[5] Kahn was one of the first to combine detailed quantitative forecasts with imaginative descriptions purportedly written by people living in the future.[6]

The narrative situates itself at the midpoint of a four-hundred year span that begins with the advent of the Industrial Revolution in England and culminates with its completion in every country of the world by the year 2176. Kahn traces key economic, demographic, resource, and environmental trends over these four centuries; and he extrapolates those trends, along with growth patterns in materials prices and availability and a host of other factors, into the distant future.[7] The basic structure of his argument is common to many futures studies, both quantitative and qualitative.

---

[5] In this regard, Kahn's work was similar to Nikolai Kondratieff's analysis of nineteenth century price behavior (including wages, interest rates, prices of raw materials, foreign trade, bank deposits, etc.). Kondratieff observed economic growth and contraction within 50-year cycles—or waves—and used the emerging patterns as a basis for predicting future economic growth.

[6] Kahn, who then worked at the RAND Corporation in Santa Monica, California, called these vignettes "scenarios," a term he reportedly adopted when nearby Hollywood studios switched to the term "screenplay."

[7] The arguments are based on assumptions that the growth of populations, economies, and other key factors that are currently expanding exponentially will begin to saturate and level off, thereby replicating on a global scale those patterns so often seen locally in the past.

Writing at a time of increasing pessimism about the world's prospects for continued economic expansion, Kahn supplied an existence proof that an adequate standard of living can eventually be provided for the entire population of the Earth. Kahn explicitly sought to influence his contemporaries' views. Worried that concerns about "limits to growth" would cause societies to slow the economic growth and technological innovation needed to fulfill the promise of the Industrial Revolution, he aimed to bolster his readers' confidence in the future. But like all narratives of the future, Kahn's work could say nothing about the implications of the many plausible paths he did not have the time or inclination to describe.

Such historical lessons are insightful and useful, but as any historian would caution, they are susceptible to many interpretations. What proves to be different about the future is likely to be as important as any similarities it has with the past. It is clear then that, at the very least, a rich collection of alternative views needs to be assembled to improve the probability that the past will be a reliable guide for decisionmaking aimed at future outcomes.

GROUP NARRATIVE PROCESSES: DELPHI AND FORESIGHT

Traditionally, narratives of the future are the work of one individual or of a collaborative team laying out a particular vision. It is clear, however, that the factors affecting the long-term future can greatly exceed the range of expertise of any small group. Thus, great interest has arisen in developing formal methodologies in which large groups of experts can combine their knowledge systematically and create narratives of the far future.

The Delphi Method Produces a Consensus-Based Response

Among the first group processes, the "Delphi" technique was developed by RAND researchers in the 1950s as a way to amalgamate expertise from a wide range of knowledge areas and divergent views and to achieve eventual consensus.[8] The Delphi process is iterative

---

[8]The earliest mention of Delphi in RAND's currently available publications is Dalkey and Helmer-Hirschberg (1962), described as an abridgment and revision of a 1951

in nature. In successive rounds, a group of experts is asked to supply responses to a list of questions. At the conclusion of every round, the participants view each other's answers and may then change their views in light of what others believe. The answers are presented anonymously to eliminate the possibility that undue weight will be placed on the responses of persons who hold particularly high status within the group.

In one early example of this approach, T. J. Gordon and Olaf Helmer (1964) led an expert panel through a series of speculations about key characteristics of the world in 1984, 2000, and beyond. Gordon and Helmer prefaced their study with disclaimers suggesting that they did not intend to predict the long-term future. Nonetheless, it is clear that they conceived their role as adding authority to predictions their policymaking audience presumably required. They described the work as driven by a desire to "lessen the chance of surprise and provide a sounder basis for long-range decisionmaking." However, anyone relying on their answers would have been surprised indeed.

Of the eight specific projections for 2000 reported in this study, seven failed to transpire as conceived by the panel.[9] Wrong guesses generated by such studies often seem humorous in retrospect, but the important thing is to recognize why the predictive task is impossible to carry out. Delphi is designed to bring a disparate group of informed opinion holders to consensus about the future, if only on ranges of probabilities. Yet, many of the topics of most interest to those organizing Delphi exercises are simply unpredictable, no matter how much is known about them. While Delphi can provide a disciplined reification of conventional wisdom, it does not provide any guarantee that the output will bear any relation to how the future unfolds.

---

document that reports one experiment in a series performed at RAND, designated internally as "Project Delphi."

[9]The predictions included the following: a world population of 5.1 billion, large-scale ocean farming and synthetic protein production, regional weather control and controlled thermonuclear power as a source of new energy, development of a universal language and "high IQ" robotic machines, mining on the Moon and a landing on Mars, weather manipulation for military purposes, and effective anti-ICBM defenses ("air-launched missiles and directed-energy beams"). The panel was closer to the mark in forecasting "general immunization against bacterial and viral diseases" though still a bit premature in forecasting the correction of hereditary defects through "molecular" engineering (pp. 40–41).

The issue of future technology development provides a good illustration. A familiar general pattern describes the entry path of many new technologies into the market (Utterback, 1994). First a period of experimentation occurs when many small companies compete with different, innovative versions of the same fundamental idea. For example, in the early development of the automobile, it was not clear whether a car was to have three wheels or four; be steered by a wheel or a tiller; be powered by internal combustion, electricity, or steam, and so forth. In the second phase, after an initial period of experimentation, a dominant design emerges. For the automobile, this occurred in the early 1920s with the steel-body, four-wheel, internal combustion-powered vehicle. Finally, the many small stakeholders coalesce into a few large firms that compete to most efficiently deliver the new product.

While this general pattern is discernible in retrospect, no panel of experts can reliably identify the ultimate winners and losers or the instances that will break the pattern. Delphi groups often identify and trace many plausible paths into the future but they cannot determine which is most likely to occur. Thus, the method errs when it encourages experts to reach consensus on the latter rather than fully articulate the former.

The use of Delphi and its derivatives has waned in the United States, but the approach continues to be employed elsewhere in the world. The Japanese government has conducted large-scale Delphi studies of expert opinion in science and technology at regular five-year intervals since 1970. And in the early 1990s, Japanese Delphi experts carried out a similar exercise jointly with Germany (NISTEP, 1994). Exercises such as these gather input from thousands of participants to cover the widest range of fields and ensure a broad canvass of expert input from each sector. This apparent strength is also a weakness because, in addition to its reliance on prediction, the Delphi method is too limited by reason of the scale of effort required to be a practical means of informing long-range policy planning.

Foresight Exercises

Unlike Delphi, which emphasizes the product of its deliberations as a principal goal, Foresight exercises focus on the deliberations themselves. The Foresight method aims to create venues where leaders from government, business, science, technology, and various other groups can come together to discuss and share both normative and positive views on future technology developments and their effects on important economic sectors and social structures. These deliberations are intended to create channels for communication as well as a better vision of what might lie over the horizon.

In the years since the United Kingdom's original exercise, Foresight-type efforts have become relatively common in Europe and Asia.[10] However, many of these exercises have adopted a substantially broader focus. In addition to technology, Foresight now also touches on social, economic, and even political issues to gain insight into trends across a broad cross section of a country's public life. Details of the method may vary, but all Foresight processes are characterized by disciplined group inquiries into the trends affecting future outcomes as well as the actions by which these trends and outcomes may be adjusted.

In practice, Foresight also struggles with the multiplicity of plausible futures. There is no fundamental reason for a Foresight exercise to be an exercise in prediction. The deep uncertainty surrounding the exploration of future possibilities in the context of Foresight represents no failure of due diligence. Rather, it is inherent in the systems that a society deems most important. Nonetheless, many Foresight participants, especially those engaged in massive efforts to canvass large numbers of individuals and communities, often share an unspoken assumption that the goal of the process is to minimize the irreducible uncertainties inherent in the forces driving toward an unknown future.

This perception may flow from the conviction that predictions are necessary precursors to effective action. Certainly, Foresight as currently practiced lacks mechanisms that can make effective use of multiple futures. The process cannot acknowledge deep uncertainty and simultaneously provide operational policy recommendations.

---

[10]This discussion does not touch on closely related efforts, such as technology roadmapping, that usually have a less grand, generally industry-specific focus than is common in Foresight exercises. The U.S. federal government does not formally engage in Foresight to inform policy choices. However, there are examples of Foresight being practiced at state and regional levels. See Ben R. Martin and John Irvine (1989).

When it achieves the one, it invariably sacrifices the other (Popper, 2002). Foresight can create alternative views of the future and thus support discussion of uncertainty, but it has no means of recommending practical strategies to address that uncertainty. To provide policy conclusions, Foresight must downplay the multiplicity of plausible futures and settle on one or a very small number of forecasts.

SIMULATION MODELING

When narrative and group processes employ quantitative data to support their visions of the far future they often rely on some type of trend analysis. That is, they extrapolate one or more technological, economic, or demographic trends on the assumption that those trends will continue into the future just as they have emerged from the past. Certainly, most physical and human processes seem highly correlated in serial fashion so that the best naive prediction is that the contours of tomorrow will be the same as those of today. Many common and successful strategies—for instance, the bureaucratic rules governing most organizations—rest on such expectations. However, shifts or discontinuities in today's trends often prove most salient in creating future dangers and opportunities and therefore determine the success policymakers have in shaping the long-term future.

Computer simulation models can play an important role in the practice of LTPA. Such models provide one of the few means to trace methodically how key components of a system will change over time as they interact with one another and, in particular, how such interactions might cause significant deviations from past trends. Simulation models generally use mathematical expressions to represent such key real-world processes as economic growth, environmental quality, and technological advances. These representations are fit to real-world data and theoretical understandings from the physical, biological, and social sciences. The simulation model can trace the evolution of the system over time, based on these representations of its parts. Thus, simulation models combine data from past trends with assumptions about the key causal relationships among relevant factors to suggest how those factors may change over time.

The World3 Simulation Model

Relatively few simulation models designed expressly for LTPA exist.[11] Those few exemplars are generally global in scope, thus reflecting the assumption that over the long term the fate of any one region of the world will depend on that of others. Some models focus on addressing specific policy questions. Others attempt to paint a more general picture of the future.[12]

The most well-known, long-term simulation model is World3, which provided the foundation for the seminal and controversial *Limits to Growth* study. Developed by Donella Meadows, Dennis Meadows, Jorgen Randers, and others in the early 1970s for the Club of Rome, an international group of businessmen, statesmen, and scientists, World3 was used to make three main arguments: current rates of economic development would outstrip the Earth's capacity to sustain human society within a century; human society could be brought into a sustainable balance; and the sooner the transition to sustainability began, the more likely it would be to succeed (Meadows and Meadows, 1972; also see Meadows, Meadows, and Randers, 1993).

World3 is a systems dynamics model, tracking long-term growth in and interactions among population, industrial capital, food production, resource consumption, and pollution.[13] Relatively simple for a

---

[11]This discussion draws from Kateryna Fonkych, "Modeling for Long-Term Policy Analysis: The Case of World3 Model," prepared for the RAND Graduate School seminar on LTPA, September 2001.

[12]Examples of simulation models with century-long spans include the International Futures (IFs) model developed in the 1990s by Barry Hughes, the Regional World model (RM) developed in the 1970s by Fred Kile and Arnold Rabehl, and the World Integrated Model (WIM) developed in 1979 by Mihajlo Mesarovic and others. Models spanning several decades, often exploring economic futures, include the GLOBUS model developed by the Science Center Berlin in the early 1980s and the FUGI econometric model developed by Akira Onishi. See Peter Brecke (1993) for a review of these global models.

[13]It includes 149 equations and tables, with 18 for population, 81 for economics, and 18 for natural resources and pollution. World3 represents all nations of the world as a single entity and simulates exponential growth in such key system components as population, economic growth and investment, resource usage and pollution, and food supply for up to a century. The model also tracks stocks of industrial capital, pollution, cultivated land, and a single-state variable that represents all of the Earth's nonrenewable natural resources.

global model, the fundamental dynamic in World3 is that economic growth engenders exponentially increasing demand for resources that eventually overwhelms the inelastic or fixed sources of supply. The world economy and population then collapse from lack of necessary resources.

World3 focuses on the interactions, called "feedbacks," among its key components—i.e., relationships among two or more variables where change in one variable causes a change in others. This, in turn, causes an additional change in the first. Positive feedbacks amplify the effects of change while negative feedbacks reduce them. Technological innovation represents a key negative feedback in World3. As environmental resources become scarce, the model allocates capital to create new technology to permit more-efficient resource use. However, the model also assumes lags that make the response less than instantaneous. Despite such technology feedbacks, four central assumptions give the World3 model a strong tendency to yield scenarios where economic growth eventually strains available resources and collapses. These assumptions are as follows:

- Growth will be exponential.

- There are natural limits on both needed resources and the sinks that absorb pollution.

- The signals society receives about impending resource scarcity, and its response to them, are distorted and delayed.

- The system's limits degrade when they are overstressed and overused.

Not surprisingly, the *Limits to Growth* study spawned a good deal of criticism (see, for example, Cole et al., 1973). That debate provides a canonical example of the weaknesses of computer simulation models for LTPA. Simulation models, despite their strengths, can still run afoul of the multiplicity of plausible long-term futures because the model-generated set of scenarios is limited by the assumptions built into the model relationships. The debate over World3—as is typical for discussions about the policy analyses stemming from simulation models—therefore focused on what relationships were included in or excluded from the model. Criticism came from many quarters, but the common theme was that the model makes assumptions that might be unwarranted and, if changed, would lead to different conclusions. In particular, the critics point to important feedbacks absent from the World3 model.

Economists criticized World3 because it fails to represent the effect of prices and markets that in the real world can work to allocate scarce resources more efficiently. As the price of diminishing natural resources rises, people will demand less, either by developing ways to use those resources more effectively or by substituting other resources. At the same time, while World3 assumes that a fixed proportion of capital goes to each of the agriculture, industry, resource extraction, and other sectors, in reality markets operate to determine how available capital is allocated among competing needs. This could avert the collapse in the model caused in part by society running out of capital to counter emerging resource constraints in different sectors.

Several critics argued that World3 incorporates fundamentally pessimistic assumptions about the rate of technological progress that do not reflect the experience of past centuries. They emphasized this point by suggesting that a late nineteenth century modeler concerned with limits to growth and using the World3 model would have predicted similar catastrophe for the coming century because no one in 1900 could have foreseen, for example, the Green Revolution. Similarly, a contemporary observer cannot predict the ramifications of today's biotechnology revolution.

Critics also argued that human society has far more foresight than was assumed in World3. In practice, societies faced with growing problems often take action to correct them. Faced with impending collapse, people might limit the rate of industrial growth or slow the rate of population growth. In short, human foresight provides a powerful feedback mechanism. World3 acknowledges such decisions and feedbacks, but it does so in a way that critics find overly simplistic and insufficient to capture the extraordinary adaptability of human society.

Structure of Simulation Models Under Deep Uncertainty

World3's shortcomings are by no means unique. The creator of any model must make choices that determine its structure, and this structure in turn determines the futures it can represent. For any computer simulation model that addresses the long-term future, no definite proof or test can determine absolutely what the structure of the model ought to be. This inconvenient fact is implied by the term "deep uncertainty." While prediction is possible for well-defined problems and isolated systems, deeply uncertain problems are subject to surprise. No finite model can contain all the potential richness of the actual, future world. Any computer simulation model used as a basis for predictive argument in support of LTPA is vulnerable to the challenge that it has neglected some factor or relationship that could prove decisive in shaping the long-term future. When a model's predictions can be repeatedly compared to actual data, its designers can appeal to this comparison to validate their selection of factors. By definition, such validation is not available for those engaged in LTPA.[14]

In recent years interest in newer simulation techniques has increased, in particular agent-based modeling, where the representations of macroscopic structure in the model can emerge from simple rule-based interactions of more microscopic entities. A classic example is Schelling's (1978) simulation, originally conducted with checkers on a board, representing the decisions of individual homeowners about where to live. This simulation showed that strongly segregated neighborhoods could emerge from very mild individual racial biases. Increased computational capabilities have enabled construction of similar bottom-up simulations that can reproduce many of the structures of modern society, such as markets (Epstein and Axtell, 1996) and states (Cederman, 1997), without specifying them a priori. Such models hold promise for LTPA because they may simulate future changes in the structure of society that the model designers did not specify or even perhaps anticipate. Nonetheless, these new simulation techniques do not escape the challenge of deep uncertainty since any macroscopic structures emerging from such

---

[14]In some cases it is informative to assume, for the sake of argument, that the future will be just like the past. It can then be possible to validate a model using historical data and to use the resulting model to make forecasts. However, such a model is only as reliable as the assumption that there will be no surprises, and the frequent failure of such assumptions limits the usefulness of this approach. (For a more extensive discussion of models that cannot be validated but are still useful, see Bankes, 1993; Bankes and Gillogly, 1994a; Hodges, 1991; Dewar et al., 1996.)

models flow from the specifications of the rules whose particulars are themselves deeply uncertain.[15]

Narratives help provide compelling visions of the future. Group narratives combine the insights of many individuals. Computer simulation models keep extrapolations consistent with the known facts about the future and help to explore the implications of a wide range of plausible futures. Yet absent a means for systematically defending the choice of factors they do and do not include under conditions of deep uncertainty, simulation models will not achieve their full potential to support LTPA because these choices must be credible to those asked to believe the simulation modeling results. To contribute to LTPA, simulation models should be utilized within a formalized decision processes that recognizes the many avenues for alternative model construction created by deep uncertainty.

FORMAL DECISION ANALYSIS UNDER CONDITIONS OF DEEP UNCERTAINTY

Humans' impressive ability to reason is nonetheless heir to a number of persistent, well-documented limitations when used to make judgments about uncertain futures (Dawes, 1998). Faced with masses of data, humans often recognize desirable trends where none exist or ignore unwanted but real patterns. Humans are notoriously bad at estimating likelihoods because they greatly overestimate their confidence in the course of future events and simultaneously hold beliefs that violate basic laws of probability. Human biases, coupled with an inability to track the implications of long causal chains, may skew judgments in ways not easily recognized. The premise of this report is that patterns of reasoning that deviate from those of formal decision analysis may often be appropriate for the real problems people face. Nonetheless, when faced with a complex and difficult problem, human beings, working alone or in groups, often convince themselves of arguments that are demonstrably not true.

---

[15]The authors have argued elsewhere that the lack of a decision analytic framework suitable for agent-based models has been a key limitation to their more widespread use for policymaking and have suggested that robust decisionmaking provides the necessary decision analytic approach (Bankes, 2002a, 2002b; Lempert, 2002a).

Traditional Decision Theory Relies on Assumptions Inappropriate for LTPA

In recent years, an explosion of public and private sector interest in scientific, quantitative methods for decisionmaking in the presence of uncertainty has occurred. The most common approach, stemming from pioneering work of the 1960s (Raiffa, 1968), is based on a *predict-then-act* paradigm that combines economic models of rational decisionmaking with methods for treating uncertainty derived from science and engineering. These formal decision-analysis methods allow the comparison of proposed decisions against normative criteria of what constitutes good and bad choices. Decisionmakers first obtain predictions of the likely outcomes from each decision option. All the potential consequences of each alternative are enumerated and then assigned a value (utility). A decision results from discovering and implementing the alternative with the most advantageous consequences.

When uncertainty is present, the consequences of alternative actions are weighted by their likelihood. The choice should then fall to the alternative that provides on average the best value (optimum expected utility). This framework has been applied successfully in a wide variety of fields, such as policy analysis, business decisionmaking, finance, and engineering. When the assumptions behind the theory hold, formal decision analysis can be shown to be the only logically consistent basis for making decisions.

These quantitative methods offer key insights but also present a dilemma when used for LTPA. They can organize vast quantities of information and ensure that any decisions based on such information are logically consistent. Yet, some key assumptions underlying traditional decision theory do not hold when one looks to the long term.[16] In particular, traditional decision theory addresses the multiplicity of plausible futures by assigning a likelihood to each and averaging the consequences. This makes sense for many short-term

---

[16]A leading decision scientist, Granger Morgan (1999), and his colleagues distinguish policy problems by the time scale and number of stakeholders involved in order to characterize the ability of traditional, predict-then-act decision analyses to address them. They find global, long-term policy problems the least amenable to traditional methods.

policy problems when some solid basis exists both for limiting the number of reasonable alternatives and for estimating their probabilities. However, probabilistic treatments do not solve the prediction problem. There is simply no way of knowing the likelihood of the key events that may shape the long-term future. A radical technological innovation or a revolutionary change in the political order could dominate future outcomes. Because experts consistently underestimate the likelihood that they will be surprised (Kahneman, Slovic, and Tversky, 1982), sophisticated elicitation of expert opinion does not solve the problem of predicting the far future.

In addition, formal decision theory seeks an optimal policy solution. That is, the ideal policy is that single option that performs "best" given some set of assumptions about the likelihood of various futures. However, such an assessment is strongly tied to the validity of the assumptions that underpin the particular analysis. Such approaches make recommendations that can be relied on only if all the assumptions made about the future turn out to be correct. Quite rightly, these are not the criteria individuals actually use when making long-term policy decisions (Rosenhead, Elton, and Gupta, 1972).

The difficulties in applying predict-then-act decision tools to LTPA arise because of the requirement for decisionmaking under conditions of deep uncertainty. Therefore, such traditional methods as decision analysis that rely on optimization for a particular future will not produce policies that have the necessary robustness to seize unexpected opportunities, adapt when things go wrong, or support the forging of consensus.

Traditional Decision Analysis Has Proven Problematic When Applied to LTPA

The problem of global climate change provides an example of the strengths and weaknesses of traditional decision analysis when applied to LTPA. Climate change is the quintessential example of a long-term policy problem. Emissions of greenhouse gases—carbon dioxide, methane, and a host of other gas molecules—trap excess heat in the atmosphere and warm the surface of the earth. Once resident in the atmosphere, these gases can persist for decades or centuries. The thermal inertia of the oceans means that today's actions can commit the earth to future warming whose effects will not become apparent for decades. In addition, the majority of anthropogenic greenhouse gases are emitted from society's capital infrastructure—power plants, factories, and transportation systems—that, once built, can operate for decades (Lempert et al., 2002). All told, decisions made today can have profound effects on greenhouse gas emissions and future environmental quality for decades and centuries. Given the inertia in the climate system and the human sources of emissions, any quantitative study of the impacts of alternative climate change policies must necessarily look a century or more into the future.

To compare alternative policies, researchers often develop "integrated assessment" computer simulation models to represent the mutually interactive behavior of the future economy and the climate system. These models relate alternative decisions to particular outcomes. For instance, a model might calculate the effects climate change could have on economies and ecosystems over the course of the twenty-first century, along with the costs of changing society's energy infrastructure in an attempt to reduce climate change. Analysts treat uncertainty about the long-term future by assigning probability distributions to such key parameters as the sensitivity of the climate system to increased greenhouse gas concentrations or the damages a changed climate might cause to human society. In the most sophisticated treatments, these distributions are elicited from leading experts in the field. Propagating these probabilities through the model gives likelihoods for each of the many plausible futures generated by the model.

Formal decision methods have yielded some important insights for the long-term problem of climate-change policymaking. They include a better understanding of the trade-offs between early and delayed action and an appreciation of the importance of efficiently allocating burden-sharing among different firms, economic sectors, and nations to lower the costs of responding to climate change. Nonetheless, the formal decision-analysis framework has important weaknesses when applied to LTPA. In particular, it assumes that the many plausible futures can be addressed by assigning a likelihood to each one and then averaging across the spectrum.[17] In the case of the far future, such probabilities, even if elicited from experts in the field, are neither much different nor much more accurate than those derived from such group processes as Delphi and Foresight.

Of course, it might be possible for a decisionmaker to become confident about a particular estimate of probabilities over a multiplicity of long-term futures. However, most LTPA challenges, including climate change, involve *many* decisionmakers whose interests, expectations, value systems, and interpretations of available scientific evidence may be widely divergent.[18] Decisionmakers and other political actors well understand that certain assumptions lend support to certain policies and that available data support a wide range of plausible futures. It therefore stands to reason that contending stakeholders will do their best to choose subjective probabilities that best support the position they wish to adopt on ideological, financial, or other grounds. Thus, when applied to LTPA, the powerful tools of formal decision analysis neither reliably provide a normatively sound basis for decisions nor, in practice, a framework that can coax consensus from multiple decisionmakers eager to champion their favorites from among a multitude of plausible futures.

SCENARIOS: MULTIPLE VIEWS OF THE FUTURES

All of the approaches discussed heretofore find their strengths tempered by their inability to confront the challenge of multiple plausible futures. In contrast, scenario-based planning is designed precisely to grapple with this multiplicity and unpredictability.

---

[17]Three key probabilities that emerge as drivers from such exercises are the following: in the absence of any policy actions, emissions of greenhouse gases will be large over the twenty-first century compared to the likelihood that they will be small; key technologies that might reduce greenhouse gas emissions, such as hydrogen-powered automobiles or carbon sequestration facilities for coal power plants, will be inexpensive rather than expensive; and the climate might experience an abrupt change in response to human interference (National Academy Press, 2002).

[18]To be sure, empirical evidence and scientific arguments can narrow the different expectations among different groups. For instance, the scientific community might generate a consensus about the subjective probability distribution for some physical parameter, such as climate sensitivity, sufficient to convince any reasonable stakeholder. For many key parameters, this is just not possible—e.g., the likelihood that new technology will substantially lower the cost of greenhouse-gas abatement.

Scenario-planning can crystallize the understanding that the long-term future may be very different from the present, and it can also help decisionmakers choose strategies based on this recognition. Even so, scenario-based planning as currently practiced does not provide a systematic foundation for comparing alternative near-term strategies that must be the ultimate goal of LTPA.

Scenario-planners conduct group processes to create narratives about the long-term future. Rather than tell a single story, the planners craft a suite of several complementary, yet fundamentally different, tales. A family of scenarios aims to span the range of plausible futures relevant to the decision at hand. In concept, such scenario families capture the fundamental truth that the future is influenced and constrained, but not determined, by the past and present. In common parlance, the word "scenario" connotes a postulated sequence of future events. Yet, in the modern practice of scenario planning, a single scenario is like a one-legged stool: it is useful only as one element of a whole that, in its entirety, spans a wide range of plausible futures.

In practice, scenario-based planning exercises lay out a small number—usually three or four—of alternative, self-consistent stories about the future. Some exercises, especially those incorporating computer simulation models in addition to detailed narratives, include calculations for dozens of cases but in the end will summarize all of those runs by reducing them to a small number of stories representing general scenario classes. Every scenario in the exercise is meant to be plausible, that is, logically self-consistent in the sense that each postulated event follows from those that come before. Some parties to the exercise may regard certain scenarios as exceedingly unlikely and undesirable, but no one should be able to prove any scenario impossible. Taken as a set, the scenarios should be compelling and acceptable to all the individuals involved.

The scenario-planning literature describes a series of steps by which a group may develop a set of scenarios. In one widely applied formulation (Schwartz, 1996), the group first identifies the decisions the scenarios are meant to inform. For instance, a firm may need to decide whether or not to invest in a new manufacturing facility. The group next lists key factors in the external environment that may affect the decision and the key driving forces that may in turn influ-

ence those factors.[19] The group then ranks the key factors and driving forces on the bases of degree of importance to the success of the
decisions and the degree of uncertainty surrounding them. The most
important driving forces and uncertainties are selected to differentiate the scenarios. Once the set of cases is so identified, the group
crafts a "scenario logic," turning each case into the outline of a story.
Then each scenario's story is fleshed out, checked for plausibility,
and given a name. Finally, the group can begin to test alternative
policies against the resulting scenarios. Some scenario-planning
exercises also identify "leading indicators" that decisionmakers can
track as they move into the future and that will give them early
warning of which scenario is coming to pass.

Scenario-Based Planning Provides a Structure for Considering Risk

By considering multiple views of the future, scenario-based planning
provides a powerful framework for organizational learning, allowing
different parts of the organization to share and understand information about risk.[20] Decisionmakers will often reject projections of the
future that deviate from what they expect or what they regard as
comfortable. Families of scenarios help overcome this barrier by
including in a package both scenarios that decisionmakers find comfortable and those that challenge them. In addition, a family of different scenarios can help groups of stakeholders with the inclination
to argue for different views of the future acknowledge and accept
that many alternative futures are plausible. Finally, families of scenarios help organizations respond more quickly to changing circumstances by providing them an opportunity to think through the
warning signs of differing futures and the responses they might make
to such signposts. In one famous example, Shell Oil responded more
nimbly to the collapse of oil prices in the early 1980s because its

---

[19]For example, the future state of the economy and customer demand for the firm's
product may affect the investment decision. These two factors may also be recognized
as being driven, at least in part, by interest rates and competitor's offerings, respectively.

[20]The "constructivist" literature on risk also stresses the importance of multiple views
of the future in transmitting and receiving information about risk to decisionmakers
and stakeholders (van Asselt, 2000).

management had thought through a set of scenarios that included such an eventuality (Schwartz, 1996).

Scenarios have been used to help think through a variety of long-term policy challenges including those centering on twenty-first century sustainability. The Special Report on Emissions Scenarios (SRES) (Nakicenovic, 1999) produced by the UN's Intergovernmental Panel on Climate Change is an example of how such efforts may be linked to simulation modeling. SRES used six different integrated assessment models to create 40 scenarios to examine the implications of a range of economic, demographic, and technological driving forces. These scenarios were gathered into four families, each representing a distinct storyline: "Rapid Convergent Growth," "Fragmented World," "Convergence with Environmental Emphasis," and "Local Sustainability." The storylines spanned two key driving forces: the degree of global integration versus regionalism and the relative regard for economic versus environmental concerns.

Each of the 40 runs of each of the six models represented a particular quantitative interpretation of one of these storylines. As a consensus report of the United Nations, the SRES scenarios were not made to yield any deeply unpleasant future scenarios. Nonetheless, the four canonical scenarios defined by the two sets of driving forces did admit to a wide range of plausible future greenhouse gas emissions paths inasmuch as they were intended to convey the uncertainty about emissions to scientists studying potential future effects of climate change and policy analysts assessing alternative policy responses.

The Global Scenarios Group (GSG) provides another important example of this genre (Raskin et al., 2002). In contrast to SRES, which sought to support policymakers but not make policy recommendations, the GSG, convened in 1995 by the Stockholm Environmental Institute, was an international group of scholars who wanted to make specific recommendations about the actions needed for a transition to sustainability. GSG created sustainability scenarios that combined detailed results of quantitative simulation modeling to capture key demographic, economic, and environmental trends with literary narratives that attempted to describe what it would be like to live through the scenarios.

To organize the myriad possibilities, the GSG laid out three general classes of scenarios and labeled them "Conventional Worlds," "Barbarization," and "Great Transition." These archetypes were intended to capture the essence of the long-term challenges facing the world today. The "Conventional Worlds" family of scenarios envisions a future of incremental change, evolving without major discontinuities and surprises, leaving human values and institutions at century's end similar to those of today. The "Barbarization" family envisions a future in which escalating crises overwhelm the capacities of today's institutions, and social, economic, and moral underpinnings collapse, plunging civilization into anarchy or tyranny. The "Great Transitions" family captures a future that develops from profound, near-term shifts in human values and institutions. Societies worldwide emphasize and achieve quality of life, material sufficiency, social solidarity, and the preservation of nature.

Figure 2.1 summarizes the trajectories of population and wealth for two representatives of each scenario family over the twenty-first century. The differing courses illustrate some of the most important quantifiable differences among the scenario groups (Gallopin et al., 1997). Within the Conventional Worlds family, the "Reference" (business-as-usual) scenario extrapolates current policies across the twenty-first century leading to great aggregate global wealth, unequally distributed, and extensive damage to the environment. The "Policy Reform" scenario explores the extent to which comprehensive and coordinated government action within the context of existing institutions and values can alleviate concerns about equity and the environment raised by the Reference scenario (Raskin et al., 1998). "Eco-Communalism" and "New Sustainability Paradigm" envision two different Great Transitions. The former embraces strong decentralization and local economic self-sufficiency. The latter represents a global civilization focused on equity and environment. The "Breakdown" scenario represents a complete collapse into small, poor, warring tribes. In "Fortress World" authoritarian rulers protect small enclaves of wealth amidst an impoverished majority.

GSG provides detailed narratives for these scenarios, called "future histories," presented in a form that might appear in a late twenty-

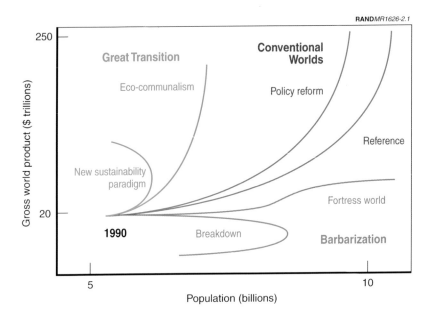

Figure 2.1—Global Trajectories for Per-Capita Income and Population in Six GSG Scenarios

first century magazine article that sought to help its readers understand the events of the early 2100s that had created their world. The GSG laid out their scenarios to argue for specific near-term actions and framed their arguments as risk-management strategies. Barbarization clearly represents a future to be avoided. Conventional Worlds, embodying current conventional wisdom, appeared superficially attractive, but the GSG argued that it was unreliable. They suggested that today's institutions and values, which balance the desire for economic growth with quality of life, may be insufficiently flexible and responsive to cope with the challenges that could appear in the twenty-first century. With luck, Conventional Worlds might endure. If it did not, Barbarization would likely result. Only Great Transition could reliably weather the storms ahead and avoid deleterious consequences. Thus, the GSG scenarios suggest that the most important step society could take to address long-term sustainable development would be near-term actions to transform the values that would shape humanity's future institutions and choices.

GSG's scenarios and similar tales provide a powerful stimulus to the imagination by helping people envision what each of the postulated futures might be like and by giving them some understanding of what driving forces might bring them about. The narrative structure of scenario families is crucial. For instance, the set of names—Barbarization, Conventional Worlds, and Great Transition—can provide a concise, memorable synopsis of the argument.[21]

Scenario-Based Planning Also Has Several Important Weaknesses for LTPA

Scenario-based planning does grapple with the multiplicity of plausible futures but has two important weaknesses for LTPA. First, the choice of any small number of scenarios to span a highly complex future is ultimately arbitrary. A scenario exercise will necessarily miss many important futures that do not make the cut into the top few. Despite best efforts, the logic used to sort the scenarios may seriously bias any conclusions drawn from them. As one small example, research in the psychology of decisionmaking indicates that humans gravitate to stories whose plot revolves around a single dramatic event rather than to those where the ending is driven by the slow accumulation of incremental change (Kahneman and Tversky, 1982). Thus, scenario-based planning exercises may make it difficult to think about responses to slowly emerging problems. Certainly they will fail to address many of the challenges that the future will hold.

Second, scenario-based planning provides no systematic means to compare alternative policy choices. Some of the literature (van der Heijden 1996) suggests means for seeking policies robust over the set

---

[21]Similarly, the "Mount Fleur" scenarios developed for South Africa in the 1980s are credited with creating a shared language for South Africans to debate the dismantling of apartheid (van der Heijden, 1996). These scenarios were popularized in public speeches and a best-selling book. The scenarios were labeled with names evocative of flight: "Ostrich" reflected an attempt by the apartheid government to conduct business as usual, "Lame Duck" envisioned a weak government overseeing a long period of transition, "Icarus" envisioned a majority government coming rapidly to power on a wave of unfulfillable promises, and "Flamingos" envisioned a black-white coalition government whose members rise slowly but fly high together. Even people who had not read about them could discuss the implications of alternative decisions using the scenario names as shorthand.

of scenarios. Dewar et al. (1993) and Dewar (2001) have developed a form of narrative scenario-based assessment called Assumption-Based Planning, which systematically finds weaknesses in existing plans and helps decisionmakers improve those plans with strategies comprising shaping actions intended to influence the future that comes to pass, hedging actions intended to reduce vulnerability if adverse futures come to pass, and signposts—observations that warn of the need to change strategies.

Although the scenario-based planning literature has systematized the process for developing scenarios, the approach for developing policy based on the results is not similarly systematic. Although best practice can incorporate scenario processes into long-range planning, most often scenario exercises by their nature stand apart as exceptional practice not formally incorporated into usual decision processes. The links making such inquiry integral to the evolving pattern of actions that must be taken to craft actual strategies, put them in place, and monitor their progress are often tenuous. More important, the largely narrative approaches used and the limitations on the ability of humans to comprehend large masses of information make it likely that scenarios with important implications for detecting opportunities, highlighting pitfalls, or informing strategic action may be missed. Scenario techniques are tremendous boons to forward-looking strategic thinking but are not formally linked to the operations of decisionmaking.

ASSESSING THE STATE OF THE ART

This brief survey suggests that traditional methods for engaging in long-term thinking and decisionmaking all grapple in different ways with the core challenge of confronting multiple, plausible futures. Taken together, these methods provide a rich foundation of wisdom on which to construct a new approach to LTPA. Narratives are unmatched in their ability to help audiences viscerally imagine what the future might be like. Delphi and Foresight exploit the fact that valuable information about the future is often best gathered through discussions among groups of informed individuals. Simulation modeling provides a quantitative structure for constructing inquires into alternatives. Decision-analysis methodology helps correct the numerous fallacies to which human reasoning is prone. Finally, scenario-based planning stresses the importance of multiple views of the future in transmitting and receiving information about uncertainties to decisionmakers and stakeholders.

However, each technique in itself proves insufficient to support the processes humans actually rely on to reason through complex and deeply uncertain issues pertaining to the long-term future. It is therefore useful to consider how best to combine the strengths of each approach while mitigating their common weakness. The guide for doing so should be a return to first principles: an explicit consideration of how the natural human mechanism for reasoning under deep uncertainty operates. Our goal is to undergird this faculty with the same level of analytic support that has been constructed for standard decision problems and, in the process, create a new way to conduct standard LTPA—or, indeed, any type of decisionmaking under conditions of deep uncertainty.

Chapter Three

ROBUST DECISIONMAKING

Often in the history of science, new tools have launched new ways of looking at the world. Galileo altered conventional views about the human place in the cosmos when he pointed his telescope, newly invented as an aid to maritime navigation, into the sky and discovered in the orbits of Jupiter's moons the existence of more than one center of motion in the universe. The microscope revealed the existence of a hitherto unsuspected world requiring new explanations. The steam engine prompted Sadi Carnot's inquiries into the intimate relationship between heat and motion, launching energy as a central concern for science and profoundly deepening the concept of the fundamental unity of nature.

Today, the computer has also become a tool poised to reshape our thinking in many fields. In particular, this technology can revolutionize the way we plan for the long-term future. The computer will not solve the intractable problems of prediction (though in some areas it may improve our ability to predict), but instead it can enable decisionmakers to ask a fundamentally different and more sensible question. Rather than first predicting the future in order to act on it, decisionmakers may now gain a systematic understanding of their best near-term options for shaping a long-term future in the absence of any reliable predictions.

This chapter introduces a new approach to LTPA based on computer-supported, robust decisionmaking methods.[1] This discussion

---

[1] Section 10.1.5 in Metz et al. (2001) surveys a variety of robust decisionmaking methods applied to the long-term policy problem of global climate change.

describes the general methods and technology used in this study and the philosophy informing its development. Chapters Four and Five demonstrate the application of a particular robust decisionmaking method to the problem of global sustainability.

DECISIONMAKING UNDER CONDITIONS OF DEEP UNCERTAINTY

The previous chapter's discussion suggests the rich legacy of insights into the practice of LTPA. The power of the modern computer now makes it possible to build on previous concepts, rendering many of them operational for the first time, to construct robust decision methods. These methods can overcome past limitations in LTPA by providing analytic support for systematic human assessment of options that accounts for the multiplicity of plausible futures. In brief, a robust decision approach allows analysts to use the computer to lay out a wide range of plausible paths into the long term. They then look for robust near-term policy options—i.e., those that when compared to the alternatives perform reasonably well across a wide range of futures using many different values to assess performance. Often strategies are robust because they are adaptive—that is, they are explicitly designed to evolve in response to new information.

Complementary Capabilities of Humans and Computers

Computer-supported robust decisionmaking at its root combines the best capabilities of humans and machines. Humans have unparalleled ability to recognize potential patterns, draw inferences, formulate new hypotheses, and intuit potential solutions to seemingly intractable problems. Humans also possess various sources of knowledge—tacit, qualitative, experiential, and pragmatic—not easily represented in traditional quantitative formalisms. Working without computers, humans can often successfully reason their way through problems of deep uncertainty, provided that their intuition about the system in question works tolerably well. They can employ narrative mental simulations to test how strategies will perform in complex situations (Kahneman and Tversky, 1982). Expert decisionmakers can use analogies from the past to suggest which strategies should work most effectively in the future (Klein, 1998). Iterative group processes can tap vast stores of relevant information. Framed around their facility for storytelling, humans can challenge each other with "what if" scenarios to probe for weaknesses in proposed plans. These processes succeed because the best response to deep uncertainty is often a strategy that, rather than being optimized for a particular predicted future, is well hedged against a variety of different futures and evolves over time as new information becomes available. Humans design successful adaptive strategies by using available information to repeatedly subject candidate strategies to examination over a range of future contingencies.

These time-tested processes can break down, however, when humans are faced with complex futures for which past experience and intuition provide an unreliable guide. Under conditions of complexity, humans rapidly lose the ability to track long causal links and the competing forces that may drive the future along one path or another.[2] Their ability to trace the "what-if" implications of proposed plans fails. When operating in organizations where individual intuition may not be easily employed or shared, humans can find it even more difficult to explore mental simulations as a group effort, even in the rare cases where many individuals share values and expectations about the future. Thus, among organizations with varying agendas or within communities that have wide-ranging interests, it becomes nearly impossible to engage in a formal commerce of ideas using these means.

Whereas humans have limited capability to process and retain information, computers excel at handling large amounts of quantitative data. They can project without error or bias the implications of those assumptions, no matter how long or complex the causal chains, and search without prejudice for counterexamples to cherished hypotheses. The process of creating computer simulation models often forces clear explication of many underlying assumptions.[3] The traditional methods of using computers to support poli-

---

[2]Klein (1998) reports that humans' mental simulations are generally quite simple. They usually contain no more than three key variables and no more than six transitions from one situation to the next.

[3]Simulation models can be crucial in helping policy analysts formalize and structure their assumptions about key driving forces. Humans generally reason with symbolic models—mental representations of real-world processes portraying cause-effect relationships among different variables based on their assumptions and observations.

cymaking required analysts to make certain key assumptions—in particular that widely agreed on models, values, and likelihoods for alternative futures exist—that are simply not true for LTPA. Thus analysts faced a Hobson's choice. They could reason with computers and reject basic tenets of the successful approaches humans use to address problems of deep uncertainty, or they could adhere to these approaches and lose the ability of computers to augment human reasoning and correct errors that often creep into human reasoning.

Building a New Approach

In recent years, the capabilities of modern computers—fast processing, often over multiple CPUs; virtually unlimited memory; and powerful, interactive visualizations—have spawned many new quantitative approaches to the ubiquitous, yet previously intractable, problems of decisionmaking under deep uncertainty. The policy analysis and integrated assessment literatures increasingly emphasize an emerging school of computational, multiscenario simulation approaches in which analysts use models to construct different scenarios (Morgan et al., 1999; van Asselt, 2000; Metz et al., 2001; Davis, 2003). Rather than aggregate the results using probabilistic weightings, this school uses the multiplicity of scenarios to make arguments from comparisons of fundamentally different, alternative cases. Statisticians and mathematicians now generate a literature on imprecise and ambiguous probabilities (de Cooman, Fine, and Seidenfeld, 2001) that allows analysts to consider the constraints of probability theory without assuming greater accuracy of information about actual probability values than is warranted.[4] The engineering

---

However, mental models permit imprecision and inconsistency. Computer models do not exhibit these shortcomings. Therefore, interaction between humans and computer models is highly complementary.

Jay Forrester (1994) has observed that "a two-way street runs between mental models and computer models." Mental models contribute much of the input for computer models. Creating a computer model requires that the mental models be clarified, unified, and extended. From the computer simulations come new insights about behavior that give new meaning to mental models. Mental models will continue to be the basis for most decisions, but those mental models can be made more relevant and more useful by interacting with computer models.

[4]The web site of the imprecise probabilities project (http://ippserv.rug.ac.be/webdb/ipp_bibliography_server.cgi) provides many papers on the topic of imprecise and ambiguous probabilities.

and operations research literatures have begun to use the criterion of robustness, as opposed to optimality, to address such situations of ambiguity (Zhou, Doyle, and Glover, 1996; Kouvelis and Yu, 1997; Ben-Haim, 2001). Advances in information technology have also resulted in the explosion of new approaches to computer simulation, such as agent-based modeling (Epstein and Axtell, 1996).

The approach to robust decisionmaking used in this report becomes feasible through the marriage of multiscenario simulation approaches with the concept of exploratory modeling (Bankes, 1993; Bankes and Gillogly, 1994; Bankes, Lempert, and Popper, 2001; Bankes, 2002b).[5] Exploratory modeling emphasizes a conceptual framework for using computer experiments to learn about the world, particularly by exploiting the interplay between computer-generated visualizations that help humans form hypotheses about properties of an ensemble of computational experiments and then conduct computer searches across that ensemble to test these hypotheses systematically. When applied to robust decision analysis, exploratory modeling uses the computer to create a large ensemble of plausible future scenarios. Here, the term "scenario" is used to represent one guess about how the world works combined with one choice among many alternative strategies people might adopt to influence outcomes. The approach then uses computer visualization and search techniques to extract information from this ensemble of scenarios that is useful in distinguishing among alternative decision options.

Robust decision methods are appropriate for many problems involving decisionmaking under conditions of deep uncertainty but are particularly powerful tools for LTPA, one of the most stressing challenges in this genre. When applied to global, long-term policy analysis, a robust decision approach aims to produce consensus on

---

[5]In "Exploratory Modeling for Policy Analysis," Bankes (1993) described what was then considered the standard approach to modeling—using the term "consolidative modeling"—as "building a model by consolidating known facts into a single package and then using it as a surrogate for the actual system." When knowledge was insufficient or other uncertainties precluded building such a surrogate, modelers were obliged to guess at details and mechanisms. "While the resulting model cannot be taken as a reliable image of the target system, it does provide a computational experiment that reveals how the world would behave if the various guesses were correct. Exploratory modeling is the use of series of such computational experiments to explore the implications of varying assumptions and hypotheses" (p. 435).

some sensible course of near-term action among the many different parties to a decision. This requires the discovery of policy options consistent with the parties' different expectations about the future and the range of values they hold.

Traditional prediction-based policy analysis pursues arguments based on an if-then framework. If the framing of the problem as a value function and a model (with its associated predictions of the future) is correct, then policymakers should take some specified action. In contrast, robust decision methods may be used to frame arguments about near-term policy actions that hold true for the full range of plausible futures and that are acknowledged as useful and valid by all concerned parties. For example, analysts might claim that a certain near-term strategy always performs well no matter what future comes to pass. An analysis seeking a robust decision would then use the computer in an attempt to falsify such hypotheses by searching for plausible futures in which the strategy would fail. The argument for the strategy is strengthened if the computer finds no future where the strategy performs poorly. The converse case weakens the argument for the strategy.

Four Key Elements for a Robust Decision Approach

Chapter Two reviewed the difficulties preventing more traditional techniques from fully addressing the needs of LTPA. A robust decision method approach would address these shortcomings by proceeding in multiple iterations as humans and computers alternately challenge each other's conclusions about futures and strategies. Four key elements or principles should govern the form and design of these interactions. These are summarized below and then described in detail in subsequent sections of this chapter:

- Consider ensembles of large numbers of scenarios. Such ensembles should contain a set of plausible futures as diverse as possible to provide a challenge set against which to test alternative near-term policies. Scenario ensembles can represent a wide range of different types of information about the long-term future. They can also facilitate group processes designed to elicit information and achieve buy-in to the analysis from stakeholders with very different values and expectations about the future.

- Seek robust, rather than optimal, strategies that perform "well enough" by meeting or exceeding selected criteria across a broad range of plausible futures and alternative ways of ranking the desirability of alternative scenarios. Robustness provides a useful criterion for LTPA because it reflects the approach many decisionmakers actually use under conditions of deep uncertainty.

- Employ adaptive strategies to achieve robustness. Adaptive strategies evolve over time in response to new information. Near-term adaptive strategies seek to influence the long-term future by shaping the options available to future decisionmakers. The near-term strategies are explicitly designed with the expectation that they will be revisited in the future.

- Design the analysis for interactive exploration of the multiplicity of plausible futures. Humans cannot track all the relevant details of the long term. Working interactively with computers, they can, however, discover and test hypotheses that prove to be true over a vast range of possibilities. Thus, computer-guided exploration of scenario and decision spaces can help humans, working individually or in groups, discover adaptive near-term strategies that are robust over large ensembles of plausible futures.

Let us now consider each key element in more detail.

CONSIDER ENSEMBLES OF SCENARIOS

To celebrate the World's Columbian Exposition in Chicago in 1893, the American Press Association asked 74 noted commentators from many fields to predict what American life would be like in the 1990s (Walter, 1992). While some writers successfully produced visions that were accurate in overall concept though rarely in detail, others amusingly extrapolated trends that have since radically changed. For instance, one essayist presciently envisioned that by the 1990s most businesses would communicate by means of electric transmissions, while another suggested that rising productivity would render three hours a long day's work. While this group included some who emerged as reasonable seers, in 1893 it was not possible to know which ones they would turn out to be.

This set of essays exemplifies LTPA's central challenge and opportunities. Any single description or model of the future is almost certainly to prove wrong. However, a diverse set of future visions may capture something important about the future that will actually transpire, though no one can identify the accurate scenarios ex ante.

Work with multiple views of the future now exists in several fields. For instance, some quantitative approaches from the machine-learning (Breiman, 1996) and weather-prediction literatures employ multiple models to improve predictions. The multiple models (generated for example from time series data) project different future states of the world that are averaged to create a single prediction frequently more accurate than that obtained by any single model (Bankes, Popper, and Lempert, 2000). Planners with specific purposes in mind also create groupings that contain small numbers of scenarios. The GSG assembled its three scenarios to argue for near-term actions to foster a Great Transition. The Intergovernmental Panel on Climate Change created its 40 SRES to identify key driving forces and characterize the range of uncertainty in future greenhouse gas emissions.[6]

Unlike the collection of 1893 essays whose only purpose was to entertain, the needs of LTPA require conscious consideration of the types of futures needed to serve a specific policy purpose. Thus, although the word "scenario" and phrase "alternative future" are generally taken as synonyms, for the purposes of LTPA it is useful to make a distinction. In this usage a "scenario" is a combination of some alternative configuration of future exogenous circumstances or relationships with the playing out of some strategic action or actions being examined.[7] The robust decision approach assembles futures as a challenge set against which to test the robustness of alternative strategies. We call any such purposefully assembled set an ensemble of scenarios (Bankes, 1993).[8]

---

[6]Earth scientists have long considered multiple views to capture uncertain information (Chamberlain, 1890).

[7]These concepts will be examined in greater detail in Chapter Four.

[8]This use is based on the meaning of ensemble as a group of complementary parts that contribute to some single effect.

A robust decision approach profits from deriving scenario ensembles that provide the greatest possible diversity of plausible futures consistent with available information. Analysts can then assess policies by testing for arguments that hold true across the ensemble. If the ensemble is sufficiently diverse, it will most likely also contain a useful approximation of the future that will actually occur. Thus, an argument true for every member of an ensemble will be true of the actual future as well.

Scenario ensembles can usefully support such arguments because humans possess much information, even about unpredictable long-term futures, that constrains the plausible members of the ensemble. Without such constraints, a completely diverse scenario ensemble would include every future that could possibly be represented and thus would support no useful arguments about policy choices. Information about the future might be in the form of quantifiable physical or economic laws—e.g., matter is conserved, current accounts must balance—or as intuitive judgments on the part of decisionmakers—e.g., the average annual rate of economic growth over the entire twenty-first century is unlikely to exceed four percent. Frequently, information about the future describes correlations among events. For instance, analysts may not be able to predict which new technology will become dominant in the decades ahead, but they can describe the well-known patterns of diffusion that any successful technology will follow (Davis, 1979; Utterback, 1994). A plausible future is one consistent with the known constraints—that is, one that cannot be refuted by any currently available information.

Probabilistic information can be incorporated with the use of ensembles in several ways. Most directly, a probability distribution can be asserted across an ensemble of scenarios that would allow the computation of the implied probabilities of outcomes. This technique is only useful when the probability distribution represents solid information. In engineering applications, probabilities based on measured frequencies can be used, in which case this technique is rigorous. Such data is not available for LTPA by definition. In principle, probability distributions can be used to represent human knowledge through subjective probabilities. In practice, subjective probabilities do not provide a natural representation for such knowledge, especially when dealing with a large number of mutually dependent uncertainties, as is the case with LTPA. Probability distributions can be a more useful construct late in an analysis, when the number of salient uncertainties has been reduced to a small number.[9]

If one does use a probability distribution as part of an LTPA, it is important to ask whether the conclusions of the analysis are robust to specification of that distribution. That is, how different would the assumed probabilities need to be to result in a different conclusion? This idea can be generalized by considering the ensemble of plausible probability distributions that can be explored using the same techniques described here for ensembles of scenarios. This is the computational equivalent to methods of imprecise probabilities now being developed by some statisticians and decision theorists.[10]

Only recently have computers become sufficiently powerful to exploit this concept of scenario ensembles. As indicated above, a completely diverse ensemble of plausible long-term futures would have a staggeringly large number of members. Fully enumerating them remains beyond present or prospective means, but modern computers can provide a useful approximation in the form of a "virtual" representation of the ensemble of all plausible futures that analysts can effectively use to inform policy choices. This representation requires two distinct types of software: a scenario generator that can create any member of the ensemble on demand and exploratory-modeling software that allows the facile generation and management of the scenario generator runs necessary to address arbitrarily complex queries about a diverse ensemble of plausible futures (Bankes, 1993).

Exploratory-Modeling Software Is the Key to the Problem of Multiple Futures

Exploratory-modeling software describes a potentially "infinite" scenario ensemble by creating a data structure that indexes every

---

[9]An example of this use of probability distributions can be found in Chapter Five.

[10]There are conceptually a variety of extensions of the basic use of ensembles described here. Any information used as input to the analysis could be systematically varied, including not only probability distributions but also constraints, functional forms, and strategies for searching ensembles as well. These more advanced applications of ensembles will not receive further consideration here.

potential member of that ensemble.[11] Any member of the ensemble may be created on demand by sending the vector of input parameters specified by the desired element in the data structure to the scenario generator. The scenario generator then conducts a run and returns the desired vector of outputs, which then completes the desired element in the data structure. Each such query may be thought of as a computational experiment that yields information about properties of the ensemble as a whole.

The specific policy arguments in each analysis define the computational experiments that need to be conducted and, hence, the sampling strategy over the ensemble. For instance, analysts might wish to test the claim that a particular strategy is a good choice by finding and characterizing futures where the strategy performs especially poorly. This effort might employ some combination of search and sensitivity-analysis algorithms with results presented to human analysts as interactive visualizations. The particular questions analysts should ask of a scenario ensemble and the algorithms and visualizations used to address those questions generally emerge during the course of the analysis and cannot be specified a priori. Thus, exploratory-modeling software should allow analysts to perform arbitrary sequences of large numbers of computational experiments, often with complex relationships to one another, in the service of advancing or refuting a specific line of policy argument.

For our research, the CARs™ computer environment was utilized to handle generation and manipulation of necessary scenarios.[12] This software provides a general environment that can encapsulate almost any type of scenario-generating software and represent (virtually) any scenario ensembles of arbitrary size and structure. The software represents each scenario as a data structure that

---

[11]This is part of the reason that ensembles are more than just sets. An underlying structure ties together the members of the ensemble, enabling their manipulation as a group as well as a purposeful rationale that originally guided the construction of any given ensemble.

[12]The Evolving Logic Computer Assisted Reasoning® system (CARs™) software provides the means for using any model or modeling techniques to run the required compound computational experiments, selecting methods for generating and examining the results of such experiments, and visualizing the output in a manner that provides integrated interaction between human operator, computational engine, and available information.

includes input parameters (representing one plausible future and one alternative strategy) and output parameters that represent the computed desirability of that scenario for today's decisionmakers measured according to one or more value systems. In addition to supporting the generations of arbitrary scenarios on demand, the software provides caching of previously run cases and services for interactively revising ensemble definitions.

Scenario Generating Software Creates the Ensemble Population

A scenario generator is a computational mechanism that can provide any scenario of the ensemble of plausible scenarios on demand. In general, they could be embodied in a wide variety of software tools— e.g., one or more computer simulation models, statistical models fit to the data, neural networks, rules capturing expert knowledge, or even an explicit look-up table relating scenario descriptors and out-comes.

While retrofitted computer simulations most frequently provide the software for scenario generation, it is important to emphasize the term "scenario generator" rather than the word "model." Scenario generators and models may look alike, but the two software artifacts are used for very different purposes.

Computer simulation models are generally used for prediction. The very word "model" bespeaks an attempt to provide an accurate por-trayal of some system. Well-established standards and statistical tools exist for validating such computer models as those used by engineers to simulate the performance of aircraft they are designing. In contrast, to properly create a challenge set for testing alternative decisions, a scenario generator should produce an ensemble with as diverse a range of plausible alternatives as possible. No widely accepted standards of rigor for assessing the quality of such scenario generators currently exist. However, it is clear that such standards should be very different from those used for predictive models. The ideal scenario generator would produce only plausible scenarios, but in constructing the software, analysts should err on the side of including potentially implausible futures. Too-early rejection of potential futures could result in making decisions that prove fragile to the reality that actually unfolds because a seemingly erroneous or initially unimportant factor may prove plausible and decisive later in the analysis. Traditional decision analysis will produce erroneous results if it relies on predictive models that produce erroneous predictions. In contrast, a robust decision method application is most vulnerable to errors from using scenario ensembles that are not sufficiently diverse. Thus, on balance, including some impossible scenarios presents less of a problem than leaving out scenarios that could prove important. During the process of analysis, scenarios that appear to be pivotal can be subjected to scrutiny and removed from the ensemble should they, on proper reflection, be deemed implausible.

During the course of analysis, humans may suggest new futures and strategies for testing and examination. It is impossible to avoid the possibility that modifications to the desired ensemble will require reimplementation of the scenario generator. However, constructing the scenario generator to produce the widest possible diversity of scenarios and using the exploratory modeling software to restrict the scenarios to be generated can clearly reduce the number of modifications required later.

Scenario generators must often produce alternative scenarios arising both from parametric uncertainty, where the values of some input variables are unknown, and from structural uncertainty, where the relationship among such factors is unknown. Exploring alternative structural relationships is more difficult to automate than exploration of parameters, but nonetheless it is often vital to respond to the questions or insights of the people involved in the analysis. As a consequence, a robust decision analysis may require varying assumed structural relationships as well as parameter values.

Scenario Ensembles Provide Best Means to Capture Information About Long-Term Futures

The flexibility inherent in the use of scenario ensembles permits amalgamation of different sources of knowledge in a way that is not easy and often not possible with models designed for prediction. If the computer simulations suggest scenarios that violate knowledge possessed by humans, such scenarios can be excluded as implausible. If humans suggest a future that requires heroic assumptions or violations of basic scientific, economic, or other principles, these otherwise undiscovered assumptions will be revealed in trying to create this future within a simulation. If humans disagree about the plausibility of scenarios, this information can be captured in sampling strategies across the ensemble. Judgments about the plausibility of futures are generally embodied both as constraints within the scenario generators—e.g., certain paths are excluded because they violate known principles of economics—and as constraints on the sampling mechanisms used to create the scenario ensemble—e.g., some parameters can never be large if certain others are small.

Scenario ensembles are powerful artifacts for eliciting information from groups and for communicating information about risk to wide audiences. Like the multiple narratives used in scenario planning, scenario ensembles offer stakeholders familiar, desired, or expected futures that make it easier for them to buy into the analysis. Ensembles also offer compelling alternative futures that can force stakeholders to question their assumptions and provide a framework to understand the views of others who might hold very different expectations about the future. In addition to its role in providing a suitable challenge set for analysis, the diversity requirement that guides the construction of scenario ensembles is crucial to building credibility among parties to a decision.

SEEK ROBUST STRATEGIES

What criterion should decisionmakers use to compare alternative solutions for long-term policy challenges? Traditional decision analysis seeks the optimal strategy, that is, the one that performs best for a fixed set of assumptions about the future. In contrast, LTPA requires a standard that allows analysts to make policy arguments true across a multiplicity of unpredictable futures.

The analysis in this report employs robustness as the appropriate criterion to assess alternative strategies in LTPA. An argument that is insensitive to significant variation in its underlying assumptions is said to be robust. In this sense, a strategy should be considered robust if it performs reasonably well compared to the alternatives across a wide range of plausible futures. The use of the normative concept of performing "well" therefore also suggests robustness should be assessed with respect to the many different value systems for assessing the performance of the strategies. When applied to LTPA, the concept of robustness has several virtues. It provides a computationally convenient basis for identifying policy arguments that are true over an ensemble of plausible futures. It offers a normative description of good choices under the conditions of deep uncertainty and multiple stakeholders that characterize LTPA. Finally, it matches the criteria decisionmakers often use when faced with such conditions.

Benefits of Robust Criteria

For many decisionmakers, robustness represents the normative criterion one ought to use when confronting deep uncertainty. When much about the future remains unknowable and many parties must concur on a path forward, those who formulate policy should seek strategies that, ideally, will not fail to minimize perceived costs or yield tangible benefits no matter which future comes to pass or who judges the success of the strategies. This concept of robustness draws from, but is not identical to, L. J. Savage's criterion of minimizing the maximum regret. Savage describes his mini-max rule as a practical rule of thumb for cases where individuals or groups are "vague" about the probabilities they attach to certain events and where all parties to the decision are certain that, for some strategies, the largest degree of loss will still seem acceptably small.[13]

Observers also suggest that many decisionmakers use the robustness criterion under conditions of deep uncertainty (Gupta and Rosenhead, 1972; Rosenhead, Elton, and Gupta, 1972). Organizations that

---

[13]In Chapter 13, Savage (1950) also describes a variety of pathologies associated with the mini-max rule. For instance, the rule often yields neither a best strategy nor a simple ordering among strategies. Furthermore, in a group context, the rule can be undemocratic because the importance of a view is independent of the number of people who hold it. Participants can easily manipulate outcomes by lying about the weights they assign to alternative futures. In some cases, the mini-max rule can be too sensitive to low-probability, high-consequence events, thereby producing clearly unreasonable results. The approach used in this study was designed to address these shortcomings by explicitly considering an interactive group process that attempts to find strategies robust across a wide range of values and expectations and that provides a systematic means of characterizing those values and expectations in which a particular choice of robust strategy will not perform well.

conduct scenario-based planning often look for strategies that perform well across their scenarios (van der Heijden, 1996). Assumption-Based Planning (Dewar et al., 1993; Dewar, 2001) has helped the U.S. Army and other organizations involved in long-term planning modify their strategies to reduce their vulnerabilities across a wide range of potential scenarios. A robustness criterion is implicit in the capital investment behavior of U.S. firms (Lempert et al., 2002). Herbert Simon's (1959) concept of "satisficing" also bears similarities to robustness. Decisionmakers engage in satisficing to seek strategies that will perform above a specified performance benchmark; those using robustness seek strategies that have some minimum threshold of performance over a variety of contingencies. Engineers implicitly use a robust criterion when they design buildings and bridges to stand in a wide range of stressing weather and earthquake conditions. Ecologists have begun to use resilience, a concept related to robustness, as a measure for how a social and economic system may successfully respond and adapt to shocks and other external changes (ICSU, 2002).[14]

Robustness is not the only possible criterion for evaluating strategies under deep uncertainty. For example, decisionmakers dissatisfied with their current status—e.g., entrepreneurs willing to risk much in pursuit of great wealth or leaders who see their nations or groups as requiring great redress for past grievance—may define failure as missing a fleeting chance to overturn the established order. Therefore, they may be far more concerned with assessing strategies suited for those few scenarios where huge gains are possible than in those many scenarios where the best they can do is maintain the status quo. To address such cases, Ben-Haim (2001) proposes an opportunity function that helps decisionmakers find and exploit favorable opportunities inherent in the uncertainty. But the audience for most long-term, global policy analyses will be those who tend to balance the desire to exploit upside opportunities with a considerable concern for avoiding downside risks. For such decisionmakers, robustness is a powerful criterion for assessing options.

---

[14]Resilience and robustness are similar concepts, though the ecological community tends to use the former to describe the properties of an ecosystem while decision analysts use the latter as a criterion to compare alternative strategies.

Implementing the Robustness Criterion

Long recognized as an important objective, robustness has not been widely used in quantitative studies because it is computationally harder to implement than optimization. In recent years, however, new approaches have begun to transcend previous limitations. Researchers at the International Centre for Integrative Studies (ICIS) have developed methods that use integrated assessment models to compare alternative management styles (represented by different choices of strategies) across different worldviews (represented by different assumptions about model parameters). They assess how the former fare when faced with a world different than the one that was assumed (van Asselt, 2000). Kouvelis and Yu (1997) have developed methods for robust discrete optimization that provide analytic mini-max solutions to a variety of problem types without any need to assign priors to alternative scenarios. Their work also draws on the scenario-planning literature for methods to help choose the scenarios that are inputs to their analysis. Researchers have also developed methods to find strategies robust against multiple priors, using formal treatments based on Bayesian analysis (Berger, 1985) and robust control theory (Zhou, Doyle, and Glover, 1996).

The robustness criterion is admirably suited to discovery of useful policy arguments that are true over a multiplicity of plausible futures. In contrast to such techniques as robust control theory or robust optimization, which place analytic constraints on the mathematics employed, iterative robust decisionmaking methods allow analysts a very wide range of choices in the mathematical representations they use in the scenario generators needed to populate the scenario ensemble. This flexible quantitative approach stems from calculating the regret of alternative strategies.

Regret is defined as the difference between the performance of a future strategy, given some value function, and that of what would have been the best performing strategy in that same future scenario. Following Savage (1950), the regret of strategy j in future f using values m is given as

$$\text{Regret}_m(j,f) = \underset{j'}{\text{Max}}\left[\text{Performance}_m(j',f)\right] - \text{Performance}_m(j,f) \qquad (3.1)$$

where strategy j' indexes through all strategies in a search to determine the one best suited to the conditions presented by future f.

A regret measure envisions people in the future benchmarking the result of a strategy against the one that would have been chosen by past decisionmakers if they had had perfect foresight. It also supports the desire of today's decisionmakers to choose strategies that, in retrospect, will not appear foolish compared to available alternatives.

It is also sometimes useful to consider the relative regret where the regret is

$$
\text{Relative\_Regret}_m(j,f) = \\
\frac{\underset{j'}{\text{Max}}\Big[\text{Performance}_m(j',f)\Big] - \text{Performance}_m(j,f)}{\underset{j'}{\text{Max}}\Big[\text{Performance}_m(j',f)\Big]}
\tag{3.2}
$$

scaled by the best performance attainable among the candidate strategies in a given scenario (Kouvelis and Yu, 1997). This measure captures the notion that decisionmakers may be more concerned with a $100 deviation from the best strategy when the best strategy earns $200 than when it earns $1,000,000.

Using either regret measure, a robust strategy is quantitatively defined as one with small regret over a wide range of plausible futures, F, using different value measures, M. This definition provides a natural basis for organizing explorations across the ensemble of plausible futures. Computer searches across the ensemble can help identify robust strategies—that is, ones with consistently small regret across many futures, F, using many value measures, M. Additional searches across the scenario ensemble can test the robustness of proposed strategies by searching for combinations of futures and values where the strategies perform poorly—that is, have high regret. Robustness based on regret also provides a good selection criterion for determining what constitutes an adequately diverse ensemble of plausible futures. The ideal ensemble would include every plausible future in which at least one of the candidate strategies might fail, according to some value system. Thus, the most interesting futures to explore and add to the ensemble are often those for which some candidate strategy has large regret.

Of course, one is rarely fortunate enough to engage in LTPA that results in an ideal strategy with good performance properties in all plausible futures judged by all relevant value systems. In practice, long-term decisionmaking becomes an exercise in juggling difficult trade-offs and in judging which values and scenarios should weigh more heavily and which should be downplayed. The choice rests, of course, on a complicated amalgam of moral, political, and goal-defined judgments.[15] The methods of LTPA being presented in this study allow decisionmakers who are confronting all types of problems to find robust strategies that reduce as much as possible the values they must balance and the wagers they must make. At the end of an LTPA exercise, they should emerge with a robust strategy and a clear understanding of the values and futures for which it performs adequately. They should also be explicitly aware of the futures and values that, by virtue of selecting the candidate strategy, have been implicitly classed as unimportant.

EMPLOY ADAPTIVE STRATEGIES

People learn. Over time, they will gain new information.[16] Accordingly, adaptive decision strategies are the means most commonly used to achieve robustness because they are designed to evolve in response to new data.[17] Faced with a multiplicity of plausible futures, a decisionmaker may settle on near-term actions but plan to adjust them in specific ways as new information renders some futures implausible and others more likely. For instance, a firm launching a new product in a test market follows an adaptive strat-

---

[15]In a robust decision analysis, these ethical and political judgments should be framed not in the abstract but in the context of the choice among specific alternative strategies. Decisionmakers are often most comfortable making such judgments precisely in such contexts (Lindblom, 1959).

[16]Sometimes, however, our perceived uncertainty increases over time if, for instance, new information causes us to question previous certainties. In addition, organizations and people sometimes forget (Benkard, 1999).

[17]Rosenhead (1989) explicitly links the ideas of robustness and adaptivity by describing them as tools to assess the flexibility achieved or denied by particular choices of near-term actions.

egy. If the product sells well, the firm expands distribution. If not, the firm may cancel or revise the offering.

In recent years, analysts concerned with decisionmaking under deep uncertainty have embraced the concept of adaptive decision strategies (Payne et al., 1993). Business publications emphasize the importance of achieving adaptivity and flexibility (Haeckel and Slywotzky, 1999). Such quantitative tools as real options analysis, which measures the value of investments in manufacturing plants and in research and development using tools first developed for valuing stock and other financial options, have allowed firms to begin to assign a value to flexibility in ways that were previously not possible (Trigeorgis, 1996). Dewar's Assumption-Based Planning process (1993, 2001) provides a framework for designing adaptive strategies. This approach comprises shaping actions intended to influence the future that comes to pass, hedging actions intended to reduce vulnerability if adverse futures come to pass, and signposts or observations that warn of the need to change the mix of actions. Increasingly, researchers are modeling the effects of alternative decisions about organizational structures for both private-sector (Carley, 2002) and public-sector policymakers (DeCanio, 2000). The environmental community has seen a surge of interest in adaptive management that also formalizes the concept of adaptive-decision strategies in the face of uncertainty (Walters, 1986).

The concept of adaptive decision strategies underlies even some of the most basic institutions of our society. Market economies and republican governments both rest on the idea that the future is unpredictable, that any human decision is prone to error, and that the social good is best served by creating and defending mechanisms that help identify and correct these errors over time. Thus, market economies encourage continual experimentation with new products and services, reward those that meet the wants of the time, and shift resources away from those producers and providers who guess wrong.[18] Many of our most fundamental freedoms, such as freedom of speech and assembly, have value as much for their instrumental

---

[18]McMillan (2002) argues that the institution of owning property is a means for enabling great flexibility in the face of unpredictable futures. For instance a biotech entrepreneur initially chooses to own her invention because no contract with a large firm could possibly anticipate all the paths the invention might take.

benefit in supporting adaptive correction to the system as a whole as for the liberty they guarantee to individuals. These freedoms ensure that governments are subject to a constant barrage of criticism that, when combined with periodic elections, helps expose and correct the state's errors. The constitutional scholar Stephen Holmes (1995, p. 179) captures this information-processing and adaptive-decision character precisely when he describes "representative government itself as a cognitive process, fashioned to maximize the production, accumulation, and implementation of politically relevant truths."

Adaptive strategies are not always appropriate, however. In some cases, decisionmakers can feign adaptivity to delay hard choices in the hopes that time will solve the problem or will pass responsibility to somebody else. On occasion, they may also find that an aura of inflexibility can force concessions from others.[19]

Nonetheless, the types of problems confronted by those who must formulate long-term policy generally demand adaptive responses. If for no other reason, decisionmakers of the future play an unavoidable role in any long-term policy process enacted in the present because their choices will ultimately determine whether or not today's strategies prove successful. The concept of adaptivity is so central to LTPA that we include it in our core definition: identifying, assessing, and choosing among near-term actions that shape options available to future generations.[20]

LTPA's very time horizon flows from this notion of adaptivity. In prediction-based policy analysis, the far horizon is often determined by how far one can accurately forecast. In LTPA, that horizon shifts to how far into the future today's actions can influence events. In some areas, such as many financial investments in a well-function-

---

[19]Thomas Schelling provides the classic example of the benefits of inflexible decisionmaking in his description of the game of "chicken" wherein two cars speed head-on toward each other down a narrow road and the winner is the driver who swerves second. Analysis should help determine when such inflexible strategies are more robust than ones that explicitly embrace flexibility.

[20]The analysis described in this report entails computational search for near-term policies that are robust to a range of plausible behaviors of future actors. This is, in effect, an application of computational game theory. Computational game theory based on computational experiments can investigate a wider range of assumptions than can mathematical game theory but cannot produce deductive proofs of general facts.

ing economy, the time value of money (as well as the internal discount rate) clearly applies and the influence of today's actions dissipates within a few years. In other areas—e.g., decisions about roads, buildings, and other capital infrastructure; the creation of new technology; decisions affecting the environment; decisions that shape the political institutions within and among nations—today's actions can have effects that reverberate for decades or centuries.[21]

In general, the most successful near-term strategies will foreclose some future options while leaving others open. For instance, parents often strive to give their children a great deal of opportunity to pursue their own interests but little chance and inclination to deviate from the parents' values. In the political arena, achieving consensus on an adaptive strategy often requires the ability to foreclose certain future options contingent on future events or information that becomes available. For example, a decision to conduct an environmental-impact assessment prior to a construction project may gather the greatest support if all parties believe the construction will go forward or not depending on the conclusions of the assessment. The proper functioning of the adaptive, error-correcting capabilities of democratic governments requires that their constitutions restrict certain behaviors, such as the ability to silence minorities or ignore the results of elections (Holmes, 1995). Similarly, market economies rely on government power to enforce private contracts entered into in anticipation of future events that may or may not unfold as envisioned. In general, economies operate in a manner that causes resources to flow to those whose risk-taking is successful and away from those whose risk-taking is not.

The most successful long-term policymakers not only work to expand their successor's capabilities. By influencing future values,

---

[21]A large body of research exists about discount rates and their impact on long-term decisionmaking. See Weitzman (1998); Cropper and Laibson (1998); and Newell and Pizer (2001) for discussions of low, hyperbolic, and uncertain discount rates, respectively. LTPA presents a multiattribute decision problem whose various elements may have different discount rates depending on their degree of substitutability. This, too, represents a factor in which preferences vary, just as they do across other aspects where preferences and value systems enter. Different discount rates would of course have very different implications in decision systems. This important point is discussed further in Chapters Four and Five. However, the example presented in Chapter Five of this report uses a common discount rate for each of the multiattribute elements.

expectations, institutions, physical endowments, and experiences, today's decisionmakers also aim to constrain the choices of future decisionmakers, and thus the paths adaptive strategies will follow as they unfold over generations.

Previous analytic approaches to LTPA have not typically supported the crafting of adaptive strategies, despite their obvious importance. Instead, those tools subtly reinforce the predisposition of analysts to think in terms of once-and-for-all policy decisions. Approaches focused on prediction do not reward adaptivity. In a predictable world, any course other than the "best" strategy is, by definition, suboptimal. On occasion, LTPA conducted with traditional decision analyses will usefully examine very simple types of adaptive strategies (Hammitt, Lempert, and Schlesinger, 1992; Manne and Richels, 1992; Nordhaus, 1994). In general, however, representing adaptive strategies in simulation models makes it computationally much more difficult, if not impossible, to use standard optimization approaches to discover good policies.

In this report, near-term strategies are explicitly presented as algorithms that initially prescribe a particular set of actions. Over time, however, the algorithms incorporate new information by monitoring one or more key trends in the internal or external environment. They may then specify new actions contingent on these observations. Often the concepts and techniques of agent-based modeling are used to represent these algorithms. Precedents exist in the psychological and organizational literatures to support representing decisionmaking in the form of algorithms. Humans generally reason using heuristics to determine which information they should track among an infinite sea of possibilities and how they should respond to the expected observations (Gigerenzer and Todd, 1999).

The design of successful adaptive decision strategies is difficult. Human policymakers often conduct mental simulations to play out the implications of decision algorithms and to appeal to experience in determining the decisionmaking heuristic they ought to use. This rule-of-thumb approach works well as long as the future is similar to the past. As the long-term future deviates from familiar experience, trusted rules of thumb can break down. To surmount this obstacle, robust decision methods compare the performance of alternative adaptive decision strategies looking for those that are robust across a large ensemble of plausible futures. These systematic explorations help decisionmakers assess alternative algorithms and choose those near-term actions that can best shape the choices available to future generations.

COMBINE MACHINE AND HUMAN CAPABILITIES INTERACTIVELY

Traditional approaches to decision support use the computer as an elaborate calculator. Analysts initially carry out detailed exercises in reasoning to draft the best model and examine all the assumptions relevant to a particular decision before feeding the results into the computer. The computer produces a ranking of alternative strategies contingent on these assumptions and, in particular, identifies the optimum. More-sophisticated efforts may also include analysis of how sensitive the optimal strategy is to the various assumptions made at the start of the analysis (Saltelli, Chan, and Scott, 2000). Then, leaving the computer aside, analysts provide a synopsis of the results to decisionmakers who engage in a subsequent and separate process to integrate the results—along with a great deal of information not included in the computer's calculations—into the institutional and political fabric of decisionmaking.[22]

Modern information technology makes possible a new and more powerful form of human-machine collaboration, one better suited to finding adaptive strategies that are robust over time. Rather than restrict the computer's province to serving as a calculator, its capabilities may be used to handle, display, and summarize vast quantities of information. The computer thus becomes a device that, from the outset of the analytical process, helps humans shape and test hypotheses about the most robust near-term actions to take in the face of multiple long-term futures. Such information is often most usefully presented in the form of interactive computer-generated

---

[22]Seen in this light, the basis for the multimillion dollar industry of selling and buying forecasts of prices, markets, political outcomes, and other future developments becomes clearer. Rather than the people buying (or selling) such information having full confidence in any particular prognostication, the value proposition that sustains this market may be that consumers find it the best available means to constantly test, modify, and add to their own internal set of plausible stories about future outcomes. They base their decisionmaking on that set of stories.

visualizations that can be regarded metaphorically as maps of the many potential paths into the long-term future. The computer helps humans identify, create, and explore the maps most useful to chart the next steps of the journey. These artifacts can become a crucial structure for a conversation among the parties to a decision. Human debates that remain unaided by data and the causal information contained in computer simulations often degenerate into contests of convincing narrative. The best-told story becomes the most compelling argument, whether or not it reflects available information. In contrast, a robust decision approach can free humans to pursue the full range of their imaginations, tethered only by the constraints of what is known.

Gathering Information

The tools used in this study allow humans to conduct a systematic, interactive, computer-aided exploration across a multiplicity of plausible futures with the goal of reaching consensus on near-term actions to shape the long-term future favorably. The particular robust decision method used in this study, RAP™[23], begins as analysts define the decision problem faced by their intended audience— e.g., government officials, business leaders, a community of stakeholders party to a decision—and gather a wide variety of information relevant to that decision. This might include quantitative data; relevant theoretical understanding from the scientific, economic, and behavioral science literatures; existing computer simulation models; existing forecasts; public positions staked out by parties to the debate; and elicitations of qualitative understandings, intuitions, and values from key parties. Analysts embody the available information in scenario generators—i.e., the computer code designed to trace out the consequences of each alternative set of assumptions.

---

[23]Evolving Logic's Robust Adaptive Planning (RAP™) method has been applied in both the public and private sectors to several previously intractable problems that required decisionmaking under conditions of complexity and deep uncertainty. Portions of the method have been used in such policy realms as science and technology planning (Lempert and Bonomo, 1998), higher education (Park and Lempert, 1998), military procurement (Brooks, Bankes, and Bennett, 1999), national security strategy (Dewar et al., 2000; Szayna et al., 2001), and environmental policy (Lempert and Schlesinger, 2000; Lempert, 2001). This report presents a first application of the method in its full form to LTPA.

As illustrated in Figure 3.1, for the purposes of LTPA the primary elements of the analysis are first divided into alternative strategies representing the different choices of near-term actions the audience for the LTPA might choose among (the policy element) and the future states of the world—the key and potentially unknown characteristics of the world that may prove important in determining the success of alternative strategies. These latter describe the elements of our current state of knowledge, including explicit statements about where the uncertainties may lie.

Forming Hypotheses About Robust Strategies

Having gathered the initial information, users next employ the computer and its scenario generator to create an ensemble of plausible scenarios. Each scenario consists of one particular choice of strat-

RAND*MR1626-3.1*

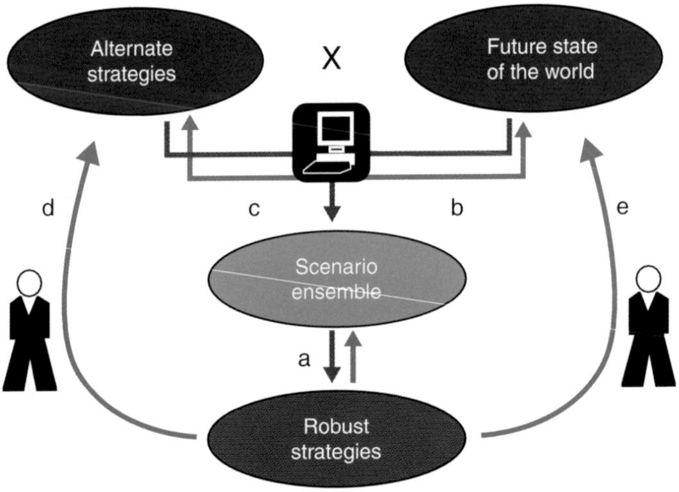

NOTE: Central lines (a, b, and c) represent computer calculations. Lines on the left and right of the figure (d and e) represent new information added by humans.

Figure 3.1—Integration of Computers and Humans in LTPA

egy, often adaptive, and one particular manifestation of a future state of the world. At this point, users can explore these scenarios by creating interactive computer visualizations referred to as "landscapes of plausible futures." These two- or three-dimensional graphics apply one or more values to compare the performance of alternative strategies across a range of futures. We have used a wide variety of such visualizations in our research (see, for instance, the review in Bankes, Lempert, and Popper, 2001). Chapter Five will employ many two-dimensional colored region plots that represent slices through the multidimensional input and output space.

Guided by the computer, users select landscapes that appear useful to address the policy questions of interest. The parameter ranges that define a particular landscape of plausible futures might be drawn from the poles of the debate over the issue at hand as well as from the analysts' intuitions about potential solutions. In addition, the choice of parameters to display on the axes of the visualizations and the values of those parameters held constant over the plot are often suggested as ones particularly interesting by computer search and sensitivity analysis. Chapter Five will demonstrate that by examining such landscapes, users can systematically explore the multiplicity of plausible futures, expand the diversity of the ensemble, detect significant patterns, and gain insight into key system drivers. Most important, by considering these ensembles of scenarios, the human-computer team then identifies strategies that are potentially robust over the range of plausible futures (path "a" in Figure 3.1)

Testing Hypotheses

Once they have formed initial claims about robust strategies, users can test and revise these hypotheses through computer search for other futures that would invalidate that claim (path "b"), and then help identify additional promising alternative strategies (path "c"). Computer-search algorithms can suggest futures that could cause candidate robust strategies to fail and find alternatives that might perform well under the newly specified conditions. Again, interactive visualization is a powerful means to help people judge whether these futures pose sufficient risks to warrant revising the strategy. The particular landscapes that display such comparisons are termed "robust regions" in harmony with the policy region analysis of Watson and Buede (1987).

These searches for "breaking" scenarios[24] may occur entirely within the space of futures and strategies contained in the initial ensemble—that is, within the capabilities of the current scenario generators. As the analysis proceeds, human participants in the analysis are encouraged to hypothesize about strategic options that might prove more robust than the current alternatives (path "d") or to suggest "surprises" that might occur in the future to cause an apparently robust strategy to fail (path "e"). These new candidates can be added to the scenario generator and their implications dispassionately explored by the computer.

Information gathering occurs throughout the analysis. Given the unpredictability of the future, no one can determine a priori all the factors that may ultimately prove relevant to the decision. Thus, users will frequently gather new information that helps define new futures relevant to the choice among strategies, that proves some futures implausible and thus removes them from the ensemble, or that represents new, potentially promising strategies.

CONCLUDING THOUGHTS

This chapter discussed the mechanics of one particular robust decision method that will be used in the analysis that follows. It is useful now to rise above the level of detail for a broader consideration of what such methods are intended to provide.

Robust decisionmaking under conditions of deep uncertainty may well fail to yield a demonstrably definitive answer of the type generally provided by traditional quantitative decision analysis. Any such exercise will surely leave futures and options unexplored. However, this iterative process does help humans and computers create a diverse ensemble of plausible futures consistent with available information and systematically identify near-term actions that are robust across the ensemble. Rather than prove conclusively that one particular strategy is the best choice, the process generates inductive

---

[24]This term will be discussed extensively in Chapter Five.

policy arguments based on a structured exploration over the multiplicity of plausible futures.

By their very nature, robust decision methods have no means to *prove* that Strategy A always performs better than Strategy B over all possible futures. Rather, they seek to make persuasive claims based on comparing the performance of those strategies over a large, well-chosen sample of scenarios. This approach stands in contrast to deductive arguments of traditional quantitative policy analysis that seeks to prove claims about the performance of alternative strategies. However useful, such methods usually achieve their rigor at a price. They require one or more highly restrictive assumptions about the representation of the problem that may well exclude a vast range of plausible futures.

The approach to combining human and machine reasoning outlined in this section is quite similar to what occurs within and among human communities discussing strategy. One person or community may argue for a decision while others attempt to discover facts or possibilities that break that argument. Similarly, one community may pose a problem; another, devise an answer. There is no guarantee that this process will produce a correct outcome, just as science provides no certainty that the next experiment will not produce unexpected results and overturn currently accepted theories.[25] Nonetheless, such competition among individuals and communities is at least one element of effective organizational decisionmaking and, indeed, the process of science itself. The goal of the present discussion is to illustrate how the computer may facilitate this powerful mode of testing and discovery and in so doing provide decisionmakers and interested communities with a different paradigm for approaching policy and strategic choice in the realm of deep uncertainty. The balance of this study will provide such an illustration.

---

[25]The parallelism stems from the recognition of how much of a role inductive reasoning plays in each. The usual simple story about the scientific process emphasizes the role of deductive approaches to falsifying hypotheses. It rarely discusses the origin of those hypotheses for testing or how the results fit into a larger frame of belief. Inductive reasoning is common in the experimental physical sciences (Popper, 1959).

Chapter Four

A FRAMEWORK FOR SCENARIO GENERATION

This chapter begins the demonstration of a robust decisionmaking approach to LTPA. This report aims not to produce specific policy recommendations but rather to reveal the new capabilities that are now available to those concerned with the long term. The analysis here involves neither the level of detail nor the level of stakeholder participation necessary for practical policy results. Its novelty rests in the framing of means for applying robust decision methods to problems of LTPA. By focusing on a specific challenge—the issue of global sustainability—the relative simplicity of the treatment that follows permits a greater focus on the main methodological themes.

An analysis of this type needs to begin with the gathering of information and the formation of the analytical framework that will sustain all that follows. This chapter will describe how this effort can be most usefully pursued to support a robust decision method LTPA.

THE CHALLENGE OF GLOBAL ENVIRONMENTAL SUSTAINABILITY

What Near-Term Strategy Will Help Ensure Strong Economic Growth and a Healthy Environment over the Course of the Twenty-First Century?

Sustainability presents a serious global policy challenge, one sufficiently novel that decisionmakers are likely to make serious mistakes if their actions are not based on quantitative policy analysis. The sustainability debate is well developed. It has identifiable camps that have generated articulate, conflicting arguments that cannot, at present, be proven to be false. A considerable store of data and computer-simulation modeling is available to inform the debate. Finally, sustainability represents a type of policy problem in which the pursuit of one widely shared goal—in this case, economic growth—can undermine other important objectives—in this case, environmental quality. Each is intrinsically valuable and necessary to the pursuit of the original goal.[1]

THE "XLRM" FRAMEWORK

As with any formal analysis, the approach used in this study requires assembling and organizing the relevant, available information. Because the process is cyclical and iterative, this step recurs throughout the course of the analysis. To help guide the process of elicitation and discovery and to serve as a formal intellectual book-keeping mechanism, it is useful to group the elements of the analysis into four categories. Policy levers ("L") are near-term actions that, in various combinations, comprise the strategies decisionmakers want to explore. Exogenous uncertainties ("X") are factors, outside the control of the decisionmakers, which may nonetheless prove important in determining the success of their strategies. In the language of scenario planning the Xs help determine the key driving forces that confront decisionmakers. Measures ("M") are the performance standards that decisionmakers and other interested communities would use to rank the desirability of various scenarios. Relationships ("R") describe the ways in which the factors relate to one another and so govern how the future may evolve over time based on the decisionmakers' choices of levers and the manifestation of the uncertainties, particularly for those attributes addressed by the measures. The relationships are represented in the scenario generator computer simulation code. As described in Chapter Three, there may often be considerable structural uncertainty about many relationships.

---

[1]Many long-range policy challenges have this character. For instance, the pursuit of physical security in the face of internal or external threats can undermine civil liberties that are both revered and necessary to support the public criticism required to correct errors in the government's security policy. In addition, the pursuit of economic growth may undermine the social stability and trust that make capitalism possible.

The primary source of information for this study was the existing literature on sustainability. However, the project team also assembled a group of RAND experts in various fields who provided methodological guidance as project advisors and acted as surrogate stakeholders from whom we could elicit opinions and information on sustainability.[2] We used the "XLRM" framework to organize our internal deliberations and to structure discussions with the advisory group.[3] In this and other work, XLRM has proven a powerful method for eliciting and organizing information relevant to decisionmaking challenges under conditions of deep uncertainty.

The remainder of this chapter provides a tour of the XLRM factors considered in the current analysis and gives additional information about how the framework can be applied to LTPA. Table 4.1 displays these factors. Those included from the outset are shown in Roman type; those added later in the process are shown in italics. In keeping with the didactic purpose of this exercise, no specific group of decisionmakers has been envisioned as having responsibility for implementing the recommendations of this study. Imagine the analysis as being placed at the disposal of such global gatherings as the United Nations environmental summits.[4]

Exogenous Uncertainties Affecting the Future (X)

Those who participate in the sustainability debate make important assumptions about the factors likely to guide the course of events over the twenty-first century. These assumptions often prove crucial to supporting their views about desirable near-term policy actions.

---

[2]The members of the panel of experts included Robert Anderson (information technology), Sandra Berry (survey techniques), James Dewar (long-term policy analysis), Robert Klitgaard (economics and development), Eric Larson (defense analysis), Julia Lowell (economics), Kevin McCarthy (demographics), David Ronfeldt (political science), and Georges Vernez (demographics).

[3]This discussion continues the long-standing practice of ordering the letters XLRM, notwithstanding the fact that it provided clearer exposition in this treatment to discuss these factors in the order found below.

[4]This work was briefed at the Science Forum, a parallel event of the World Summit on Sustainable Development held in Johannesburg, South Africa, on August 31, 2002 (Lempert, 2002b).

Table 4.1

Key Factors Used to Construct Ensembles of Sustainability Scenarios

(X) Exogenous Uncertainties	(L) Policy Levers
Economy	Policies to speed decoupling rate
Growth rates	*Near-term milestones*
Decoupling rates	*Ability to relax near-term milestones*
Effect of environmental degradation on growth	
Cost and effectiveness of policy interventions	
Coupling between North and South	
Potential surprises	
Environment	
Effect of pollution on environmental carrying capacity	
Resiliency of environment	
Potential surprises	
Demographics	
Population trajectories	
Effect of economic growth and environmental quality on population	
Potential surprises	
Future Decisionmakers	
Information	
Values	
Capabilities	
Potential surprises	

(R) Relationships	(M) Measures for Ranking Scenarios
Equations contained in modified "Wonderland" scenario generator	Rate of improvement in
	GDP per capita
	Longevity
	Environmental quality
	Weightings
	North versus South[a]
	Environmental versus nonenvironmental measures
	Discount rate
	Vantage years

[a]"North" here is shorthand for Organization for Economic Cooperation and Development (OECD) countries and "South" is shorthand for countries that are not members of the OECD.

The project team reviewed the sustainability literature and extracted what appeared to be the key, sometimes implicit, assumptions underlying the proposals of the various camps and then, as shown in the upper left quadrant of Table 4.1, divided them into four categories: those affecting the economy, environment, demographics, and behavior of future decisionmakers. This section will review these exogenous uncertainties and describe how they were represented in the computer to conduct the LTPA.

Sustainability has become a key agenda item for many businesses, governments at all levels, and for a wide variety of nongovernmental organizations and other citizen's groups. At one extreme of the sustainability debate lie those who believe that catastrophe awaits if society fails to take immediate and aggressive steps to reduce pollution. Perhaps the most famous exemplars of this camp are the "Limits to Growth" advocates, such as Paul Ehrlich and the Club of Rome (Meadows and Meadows, 1972). Noting the properties of exponential growth and the astounding fact that up to 40 percent of all the energy in the Earth's biosphere is currently consumed by humans (Vitousek et al., 1986), these commentators argue that the environment imposes real physical limits on the growth of both human populations and economic activity. Other commentators such as Bill McKibbon (1989), following in the tradition of Aldo Leopold (1949), focus on moral rather than physical constraints. They assert that human activities are threatening to engulf all of nature, thereby creating a sterile planet where nothing lives that is unmanaged by humans. They question whether material possessions are worth such a cost.

At the other extreme are those such as Julian Simon (Myers, Myers, and Simon, 1994) and, more recently, Bjorn Lomborg (2001), who argue that human ingenuity mobilized by scarcity can eliminate almost any credible environmental constraint. They note that, over the decades, the forecast price increases of most resources deemed "nonrenew-able" have failed to materialize because the incentives and technologies for extraction or for substitution have always improved at a faster rate than society's demand has increased. The real price of oil, for example, is now lower on average than it has ever been except when shortages induced by political crisis or war temporarily inflate the price. The "Unlimited Growth" advocates recommend faster economic growth as a means of alleviating poverty and a host of other human miseries. Many in this camp take a libertarian bent, focusing on human ingenuity that draws its incentives from markets. Others, such as Gregg Easterbrook (1995), argue that the dramatic improvement in environmental quality in developed countries is the most significant, unsung success story of the last three decades.[5] However, they credit innovations in government institutions and regulations that give expression to new human values as the prime movers of this promising trend.

Of course, most commentators and policymakers lie between these two extremes. As commonly conceived, sustainable development seeks to eliminate the seemingly irreconcilable conflicts between "Limits To" and "Unlimited" Growth. As first expressed by the Bruntland Commission, sustainability is often defined as meeting the needs of the present while not compromising the ability of future generations to meet their needs. A vast body of literature attempts to help governments and businesses balance the goals of economic growth and environmental protection. For example, in the GSG scenarios described in Chapter Two, the "Conventional Worlds" scenarios capture the worldview of those who believe that well-regulated markets as currently constituted can meet the challenges of sustainable development. "Barbarization" scenarios represent those who believe that future challenges may cause such institutions to fail. "Great Transition" offers an alternative set of near-term policies designed to help humankind avoid the potential dangers some see in our current path.

The key economic assumptions that distinguish the positions of the participants in the sustainability debate include the exogenous rates of economic growth—i.e., what the base growth rates will be in the absence of environmental degradation and policy designed to prevent such degradation—and the decoupling rate (Azar, Holmberg, and Karlsson, 2002)—that is, that rate at which technological and other forms of innovation reduce the amount of pollution generated per unit of economic output. Some of the most important differences between the Limits to Growth and Unlimited Growth camps rest on assumptions about whether or not the future decoupling rate will outpace the future rate of economic growth. Other core

---

[5]The story is often different in the developing countries of the Southern Hemisphere.

assumptions include the effect of any environmental degradation on economic growth; the costs, effectiveness, and time lags associated with any policy interventions; and the extent to which changes in patterns of growth in either the North or the South affect growth in the other region. While much information exists that can inform judgments about each of these factors, they cannot be predicted with any certainty even a decade, much less a century or more, into the future.

Key environmental assumptions include the effect of annual levels of pollution on environmental quality, particularly whether and at what level there exist any critical pollution thresholds beyond which emerge abrupt, discontinuous changes in environmental damage.[6] In addition, different parties have different views of the resilience of the environment under stress. Some see the environment as fragile (e.g., push it too far and it will collapse) while others see it as fundamentally resilient (e.g., even if pollution destroys the most vulnerable ecosystems, the remaining ones will be more difficult to damage).

Key demographic assumptions include trends in birth and death rates in different regions of the world,[7] how these trends would be affected by changes in wealth per capita, and the effect of environmental degradation on birth and death rates.

Finally, the parties to the sustainability debate often hold radically different views about the capabilities of future decisionmakers to detect and successfully respond to the environmental problems this generation bequeaths to them. One of the crucial differences between the GSG's "Conventional Worlds" and "Barbarization" scenarios is the set of assumptions about the ability of future generations to respond to increasing environmental stress. In the former scenarios, far-sighted decisionmakers pay attention to warning signs and adapt to emerging challenges. In the latter, society cannot summon the will to respond in time. In fact, a strong theme of moral equivalency among generations underlies many of the arguments of those who favor near-term actions to address sustainability con-

_____

[6]There is, for instance, increasing interest in and understanding of potential abrupt changes in the climate system that might be caused by human actions. See National Academy Press (2002).

[7]Migration rates are also important, but not included in Table 4.1.

cerns. Very much in the spirit of the adaptive decisionmakers described in Chapter Three who attempt to shape the values and experience of those in the future, these parties assert that future generations will need to respond wisely to early warning signs of environmental dangers and that one of the most important legacies this generation can bequeath to its descendants is the example of its having taken such actions.

Near-Term Policy Levers (L)

In the face of deep uncertainty, potential actions nevertheless could be taken in the present to decisively shape the long-term future. The sustainability literature is replete with suggestions for policies ranging from environmental regulations to enhancing innovation, changing values, creating institutions, or conducting environmental research. The policy community is struggling with strategies for combining these potential actions into coherent and effective implementation plans.

A key simplification in this demonstration analysis, and the primary reason it cannot support actionable policy conclusions, is that it considers a greatly truncated menu of policy levers. In fact, the analysis begins by presuming that only one option is available: a pollution tax that can speed the decoupling rate—that is, the rate at which innovation reduces the pollution society generates per unit of economic output. Any chosen lever will have associated costs and time lags. The actual values of these factors are among the key uncertainties. This analysis takes a global perspective and thus does not consider questions of a game-theoretic character regarding how policy choices in one region might affect the choices made in another, even though it is clear that regions are linked economically and environmentally.

In later stages of the analysis, we introduce an important, though still highly limited, addition to the menu of policy levers. Decisionmakers may set near-term milestones for the rate at which emissions are allowed to grow. As described in detail in Chapter Five, once a milestone is set, the prevailing assumption is that society then adjusts its policies to achieve an innovation rate that will allow attainment of whatever emission-intensity reduction level is required to meet the milestones. Such milestones are of much current interest to policymakers. For example, this is the philosophy that informs the effort to frame and meet the UN Millennium Development Goals. Our project might well be characterized as one designed to develop hitherto unavailable quantitative means for assessing and choosing such milestones.

A simplified set of policy levers allowed us to focus on developing straightforward (though complex) methodology to apply to much more complicated cases than we used in our demonstration. It also yields what appears to be a surprisingly interesting policy story. It is important to note, however, that the methodology itself is completely general in the types of uncertainties, levers, relationships, and measures that may be considered and, thus, should be applicable to a very wide range of policy problems.

Measures for Ranking Scenarios (M)

In addition to their vastly different expectations about the future, many of the parties to the global sustainability debate also have fundamentally different value systems. Citizens of wealthy or developing nations, environmentalists, industrialists, and public officials may hold decidedly contrary views about what constitutes a desirable long-term future. Any attempt to compare the efficacy of alternative near-term policies must not only be robust across many future scenarios, but it must also accommodate a range of viewpoints regarding desirable outcomes.

In the language of decision analysis, the challenge of global environmental sustainability is unavoidably a multiattribute problem (Keeney, 1976). This goes beyond issues of subjective taste or differing utilities. Even an internally consistent set of desires may generate a number of success measures that may not be completely synonymous. For example, a business firm might consider such indicators of success as profitability, market share, and shareholder asset value to be equally important, to say nothing of measures that might arise out of the particular nature of corporate culture. Therefore, the realm of measurement must be included in formal explorations of LTPA.

An extensive literature exists on the proper way to measure sustainability and the betterment of the human condition. Many alternatives have been proposed to serve a variety of purposes. Standard economic national-income accounting measures gross domestic product, current account balances, and a family of related concepts. Although they have few competitors for commonality of use, such metrics were not designed or intended to address fundamental concepts such as human development and environmental soundness. To remedy some of the perceived shortcomings, the UN Development Programme (UNDP) annually issues the Human Development Index (HDI) for 162 countries (UNDP, 1990, 2001). The HDI aims to summarize and compare trends in development among countries rather than to comprehensively measure the progress in individual nations. The index for any country is a weighted sum of its income per capita and levels of education and longevity.[8] Many other summary and comparative measures have also been proposed to focus on additional or different measures of the human condition. For instance, since 1972 Freedom House has published its annual Freedom Index ranking countries according to the degree they respect human rights and are democratically governed (available at http://www.freedomhouse.org). There are also ongoing, though as yet unadopted, efforts to define a "Green GDP" to account for changes in environmental quality as well as economic growth (Nordhaus and Kokkelenberg, 1999).[9] These Green GDP measures often account for natural resource depletion and pollution levels. They do not, however, generally address the health of ecosystems and their ability to deliver important services to society, also potentially critical components of a healthy environment.

In this report, evaluation of the desirability of alternative scenarios is performed by using a family of measures modeled after both the HDI and the concept of a Green GDP. This choice was driven by the need to use measures that could be compiled from computer simulation output derived from our simple scenario generator. The desire was not to take a particular stance regarding which of the many proposals

---

[8]The UNDP describes the HDI as measuring "the overall achievements in three basic dimensions of human development: longevity, knowledge, and a decent standard of living. It is measured by life expectancy, educational attainment (adult literacy combined with primary, secondary, and tertiary enrollment), and adjusted income per capita in purchasing power parity (PPP) U.S. dollars."

[9]See Prescott-Allen (2001) for an attempt to provide a comprehensive set of measures for human and ecological well-being.

for measuring the future human condition are most appropriate. Rather, the point was to make explicit the need to use multiple measures of the relative desirability of alternative scenarios. For an analysis in an actual decision-support role intended to produce policy-relevant conclusions, it would be necessary to craft measures that better capture the full range of aspirations held by all of the individuals and groups who must participate in any near-term decisions about long-term sustainability. The treatment here merely suggests measures that appear to characterize key points of view represented in the sustainability literature. While the selected measures cannot be claimed to represent the values of actual individuals, they do support a demonstration of how sets of such measures can be used in LTPA.

The measures used in this report focus on three key time series produced by the scenario generator: output per capita, longevity, and environmental carrying capacity. The first two are represented in the HDI, albeit in somewhat different forms. The third time series represents one plausible component often proposed for a Green GDP. A series for level of educational attainment could not also be included because the scenario generator used in this robust decision analysis contains no module for education. The environmental carrying capacity considered here is clearly an abstraction that serves to side step important problems of data availability and interpretation confronting those currently trying to construct comprehensive measures of environmental quality.

Appendix A fully describes the measures used in this report to rank the desirability of various scenarios. In brief, however, each measure summarizes the average rate of annual change exhibited by two or three time series for a century or more into the future. These average rates of change are discounted to the present so that near-term improvements (or degradations) in output per capita, longevity, or environmental carrying capacity count more heavily than those in the far future. Many economically focused measures of the human condition use such rate-based approaches. The HDI does not, focusing instead on annual snapshots of the relative performance of a large number of countries. The rate-of-change measure is preferable for this study, which must assess trajectories for a small number of regions over a large number of years.

This approach yielded four distinct measures of the desirability of alternative scenario outcomes from the discounted time series by applying different weightings across them. As shown in Table 4.2, the weightings differ in the emphases they place on the balance between outcomes in the developed "North" and the developing "South" and between the economic and environmental time series. The first measure, called the "North Quasi-HDI," considers only output per capita and longevity in the North. The second, "World Quasi-HDI," uses a population-based weighting for the output per capita and longevity in both the North and South in aggregate. The third, "North Green-HDI," again focuses on the North only but balances losses in environmental carrying capacity by weighting them equally with any gains in output per capita and longevity. The fourth, "World Green-HDI," uses the same weighting scheme population weighted for the entire world.[10]

The best way to get a sense of how these measures perform is to apply them to past epochs. Table 4.3 shows a World Quasi-HDI ranking for several historical periods using a 2 percent discount rate. The long-lasting U.S. economic boom in the late twentieth century (post-1950) would score 3 percent annualized growth in the quasi-HDI measure while the twentieth century in its entirety would rank a 2.1 percent, with rapid growth at the beginning and end balanced by

Table 4.2

Four Measures Used to Assess Ensemble of Sustainability Scenarios

	North	World
Quasi-HDI	N$, includes North GDP/capita and longevity	W$, includes Global GDP/capita and longevity
Green-HDI	NG, includes North GDP/capita, longevity, and environmental carrying capacity	WG, includes Global GDP/capita, longevity, and environmental carrying capacity

[10]One could create additional measures by using different discount rates and/or different vantage years—that is, the year at which the time series are discounted to. Variations in discount rate addresses different weightings of the far-term versus the near-term future and variations in vantage year describe how people 25 or 50 years in the future might rank various scenarios. This report does not cover such explorations.

Table 4.3

World Quasi-HDI Measure Applied to
Past Centuries

Period	World Quasi-HDI
1800 to 2000	1.2%
1900 to 2000	2.1%
1950 to 2000	3.0%

two World Wars and a Depression in the middle. Not shown independently in the table, the decades of the 1930s and 1940s would score less than 1 percent. Although this illustration is for only one of the measures, the four alternatives were scaled so that each produces a similar range of numerical values. Thus, Table 4.3 suggests that a score of 3 percent using any of the four measures is an excellent outcome while a score of 1 percent is bleak, given the experience of recent decades. Calculating the "regrets" of alternative strategies (as required to assess robustness) focuses on differences among scores. Table 4.3 suggests that a regret of 1 percent or more (roughly the difference between a score of 2.1 percent and the score of less than 1 percent for the years of Depression and World War) would mean the equivalent of exchanging the performance of the twentieth century as a whole for that of the 1930s and 1940s stretched across a full 100 years.

Ultimately, the emphasis placed on measures is crucial for LTPA where a homogeneous set of values by which one might measure the desirability of outcomes may not be assumed across time or even in the contemporary setting across interest communities. Just as the uncertainties, levers, and alternative relationships must be fully investigated, so too must the metrics and values of assessment be made an explicit part of the analytical exploration required for LTPA.

Relationships in the Scenario Generator (R)

In the XLRM framework, the Rs—or relationships—signify the links among the Xs and Ls as inputs or descriptors of a scenario and the Ms that measure the relative desirability of alternative scenarios. These linkages define the steps and mathematical calculations carried out by the scenario generator computer software. The methodological focus of this research effort argued against creating a scenario generator from scratch, even though the design requirements for writing the necessary code are different from those of computer models intended for prediction. It seemed more prudent to take a model from the existing literature and use it to frame the scenario-generation effort because our goal was to emphasize the overall method and avoid inappropriate concentration on modeling details.

We therefore conducted a survey of computer models relevant to global sustainability. Several offered significantly more detail and were more widely utilized than the one ultimately selected. However, none could produce a sufficiently wide range of scenarios to effectively address the issue of robustness. In general, a model designed for prediction will strive for validity through as precise as possible a representation of particular phenomenology. Thus predictive models may often prove ill-suited to producing the necessary diversity of scenarios and will require important modifications to serve as scenario generators. The more complicated models considered for use as scenario generators would have been much more difficult to modify because of their complexity, the lack of access to source code, or both. Consequently, they would not have offered an acceptable starting point for representing crucial aspects of the robust decision approach—e.g., consideration of near-term adaptive policies and the adaptive responses of future generations. As a consequence of these considerations, a simple systems-dynamics model known as "Wonderland," based on the work of Herbert and Leeves (1998), was chosen as the scenario generator for this analysis.

The Wonderland Model. Wonderland tracks changes in the economy, demographics, and environment. The original one-region model was modified for this project to represent a two-region world consisting of the current OECD (Organization for Economic Cooperation and Development, that is, the North) and non-OECD (South) countries. The economic module tracks GDP per capita. The demographic module tracks birth and death rates and total population. The environmental module tracks an abstract construct reinterpreted for this work as a natural "carrying capacity" that can be degraded when pollution exceeds some threshold level. Neither the threshold level nor the damage sustained by the environment is observable until the threshold is actually reached, and the carrying capacity then declines as a result. In Wonderland, growing population and wealth can increase annual pollution. Innovation of various types can speed the rate of decoupling—that is, the rate at which the pollution generated per unit economic output decreases. The net change in pollution depends on competition between these trends.

The scenario generator derived from Wonderland has 43 input parameters that embrace the XLRM framework's key uncertainties. Parameter values are constrained by requiring the model to reproduce past trends as well as a range of future forecasts of economic growth, population, and environmental performance. The details of the scenario generator and its parameters are found in Appendix A.

When conducting LTPA, it is crucial to distinguish between policy actions taken in the present and those potentially available to future generations. To capture this distinction, a second modification to Wonderland represents sustainability policy in the twenty-first century as a two-period decision problem as shown in Figure 4.1. In the near-term, decisionmakers must formulate policies even when they lack solid information on the extent of the sustainability problem— i.e., they do not know the actual values of any of the exogenous

RAND*MR1626-4.1*

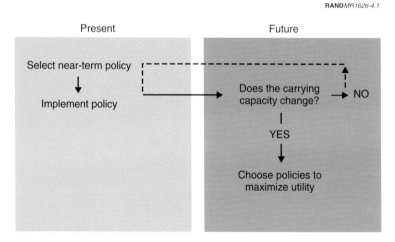

NOTE: Numerical values and equations describing the strategies are given in Table B.2.

Figure 4.1—Two-Period Decision with Fixed Near-Term Strategy

uncertainties. At some future date, those uncertain values will be revealed if and when society observes a change in the environmental carrying capacity as described above. Observation of actual damage then signals the beginning of the second, future decision period. At that time, decisionmakers will choose a strategy that maximizes their utility, given their then-current circumstances and information.

At the beginning of the first period, near-term decisionmakers choose policies that will accelerate the exogenous rate of innovation by some fixed amount. This policy remains in force until future decisionmakers detect a change in environmental carrying capacity. At this point, we assume future decisionmakers have sufficient information to choose policies based on their expected utility as described below.

The modified Wonderland scenario generator represents future decisionmakers as choosing a strategy based on one or both of two policy levers. They can engage in policies designed to speed decoupling rates, and they can make investments intended to reduce their society's vulnerability to decline in carrying capacity. This second option may reduce the need for future decisionmakers to pay for large increases in the decoupling rate. This second option is important to consider in this analysis because scenarios in which future decisionmakers choose it may be valued quite differently by the various parties to today's sustainability debate. For instance, those greatly concerned with the natural environment would be aghast to view a future where people save themselves but allow the environment to perish. Those who value only the more traditional economic measures of well-being might find such a scenario acceptable. The parameters describing future decisionmakers include their ability to detect changes in carrying capacity, the exponents of their utility functions, the relative weight they place on economic growth versus environmental quality, their discount rates, and the costs and effectiveness of their adaptation policies. These factors are all treated as exogenous uncertainties within the XLRM framework.

There is a long list of reasons why runs of the Wonderland computer code cannot be construed as reliable forecasts of the future, but forecasting is not our intent. To be valid for our purposes, a scenario generator must be able to represent the crucial aspects of the problem as captured by the XLRM framework and it must generate a sufficient diversity of plausible worlds to challenge candidate robust strategies. As will be discussed in subsequent chapters, the modified Wonderland scenario generator does well with the second criterion but is deficient in the first.

It is important to recognize that, while Wonderland's lack of detail seriously affects the policies it can compare, the model's simplicity has nothing to do with its inability to predict the long-term future. Too often analysts and policymakers demand increasingly complicated models in a vain attempt to increase the predictive power of their models.[11] Adding detail to a model can usefully admit more information into the analysis, some of which may help establish certain scenarios as implausible or prove important to comparisons among alternative strategies. Additional detail may also permit more-insightful examination of important policy levers. For instance, this study was unable to consider how investments in education or scientific research might add to a robust near-term sustainability strategy because Wonderland does not include either factor. But complicating a model as a means of reducing uncertainty is often like a dog chasing its tail. No matter how many times analysts further elaborate their models, no additional level of detail can ever enable a model to reliably predict an unpredictable future.

Wonderland suits the purposes of this study because of its ability, given a sufficiently wide range of input parameter values, to generate the range of outcomes frequently observed in scenario-based thinking about the future of the planet. Its simplicity and publicly available source code permit ease of modification and exploration over alternative mathematical relationships between the uncertainties, levers, and measures. Wonderland has many obvious deficiencies. To the extent that its simplicity makes abundantly clear that prediction is not its goal, Wonderland's manifest shortcomings can, for this study, also serve as virtues.

---

[11]The costs of more-complicated models include the increased resources needed to construct them and the decreased ability of analysts and decisionmakers to understand why they behave the way they do.

Chapter Five

IMPLEMENTING ROBUST DECISIONMAKING

The preceding chapter described a framework (XLRM) and a scenario generator, and it explained how the wide variety of information contained in them defines a multiplicity of plausible long-term futures relevant to sustainable development. The present chapter applies the approach described in Chapter Three to the problem of global sustainability and shows how this information may be applied to identify society's best near-term strategies for shaping this future to achieve positive outcomes. It should not be a surprise that computer simulation model(s) using a wide range of inputs can trace a multitude of paths into the long-term future. The challenge is using this information to support useful policy arguments.

Both to demonstrate clearly a robust decision strategy for conducting LTPA and to provide a detailed explanation of the process, this chapter is organized into two principal parts. The first describes the overall "story" of the analysis, emphasizing the methodology. Subsequent sections provide details of each step in that analysis.

OVERVIEW: INTERACTIVE ANALYSIS OF SUSTAINABLE DEVELOPMENT

Chapter Three described an iterative procedure in which robust decision methods (RAP™, in particular,) supported by modern computation, create an analytic methodology that combines the best features of narrative scenario-based planning, group processes, simulation modeling, and quantitative decision analysis. This new approach is characterized by the four key elements of LTPA, as enumerated in Chapter Three: ensembles, robustness, adaptivity, and interactive exploration.

Each of these four elements will now be demonstrated in the context of the problem of sustainable development following the procedure shown in Figure 3.1. Because the illustrations in this chapter contain icons indicating at what point in that process diagram each one was created, the reader may find it useful to refer to Figure 3.1 during the following description.

This chapter emphasizes the iterative nature of the robust decision analysis process. The explication will contrast strongly with the discussions that usually stem from traditional analytic approaches that tend to conceal the actual "sausage making" with its inevitable false starts and struggles. While the traditional approach is certainly appropriate for many types of decision problems, quantitative LTPA requires frequent and systematic revisiting of assumptions because, no matter how clever an analysis of long-term policy may be, there is never a final conclusion and no end of the need for subjecting the current best ideas to continual stress testing.

The analytic process begins by defining initial ensembles of alternative strategies and future states of the world. The cross product of these two sets forms the scenario ensemble, defined by all the possible combinations of exogenous uncertainties, levers, and configurations of the scenario generator(s) shown in Table 4.1.

Once these ensembles have been defined, and the needed scenario generator(s) created, surveying the ensemble of potential scenarios becomes possible. The next step, then, is to conduct such a survey, with a focus on discovering strategies with properties that suggest their promise in meeting long-term challenges. We are particularly interested in strategies that are robust across broad ranges of different futures and alternative value systems. This step is marked "a" in Figure 3.1.

The next section describes a landscape of cases created in this fashion. With users considering alternative values of a few key uncertainties while keeping other indices into the ensemble of plausible futures at their nominal values, one or more strategies may emerge that are robust over a wide range of futures. The candidate strategy/strategies will each increase by some fixed amount the rate of decoupling—that is, the rate at which the environmental impact of a unit of economic activity is reduced. The node labeled "Robust Strategies" in Figure 3.1 indicates the discovery of such candidate strateg(ies).

In the traditional decision analysis approach to this problem, the determination of the "optimal" improvement in decoupling rate would be viewed as a conclusion. This conclusion would, however, be a prime target for any stakeholders who objected to its implications and could find fault (often very easily accomplished) with the assumptions embedded in the analysis. By contrast, viewed as the initial result in a robust decision exercise, a strategy is merely a starting point for stress testing and evolving a more insightful stance to formulating strategy. From the "Robust Strategies" node, the analysis may take multiple paths, depending on whether or not a candidate robust policy has been discovered. The first pass through the diagram produces one or more candidate robust strategies. Consequently, the focus of the next stage of the analysis is to attempt to break a candidate strategy by finding a future (or a relevant value system) for which it performs too poorly to be acceptable. In the chapter's third section, this is accomplished by means of a more thorough search through the ensemble of plausible futures (path "b" in Figure 3.1). That search (mediated in this case by human analysts exploring alternative futures) reveals in particular that the candidate strategy does not perform satisfactorily in many futures for value systems sensitive to environmental concerns or the fate of the less-developed regions of the planet (the "South").

The success in breaking the previous candidate strategy returns us to the bottom of the figure. But this time there exists no robust strategy, and so the analysis moves along path "c," searching for a new strategy that will be robust across all futures discovered to be important to this point. If that search finds a new robust candidate, the process iterates, seeking futures to break the strategy, and so forth. This time, however, exploration demonstrates that there is no policy-driven increase in the decoupling rate that performs acceptably for all futures. As the scenario generator was initially constructed to examine only this one type of strategy, the failure presents a problem common to studies based on simulation models.

When preparing for LTPA, it is desirable to build as much flexibility into the scenario generator as possible so that changes to the ensemble of futures or strategies can be made without changes to the software. However, given the unpredictability of the long-term future, it will always be possible to exhaust the ability of the scenario generator to create the cases users will wish to examine. In such instances, direct intervention by humans to revise the tools of the analysis can become necessary. Thus, the full interactive methodology allows for the human users to modify the ensembles and pursue the analysis wherever its insights lead. These are the arcs "d" and "e" in Figure 3.1, which represent human augmentation of the framework for the analysis, modifying as needed the ensemble of plausible scenarios or the universe of possible strategies.

In this demonstration analysis, once it has been determined that no increase in decoupling rate suffices to provide a robust strategy, it is up to human ingenuity to devise an elaboration of the ensemble of strategies to allow for greater robustness (arc "d"). For long-term policy analysis, adaptivity becomes a vital tool. The fourth section describes enhancement of the ensemble's strategies to include a class of adaptive strategies based on setting milestones. This expansion of the range of policies is a product of human creativity, stimulated by the results of computational experiments done to this point. The computer can be used, through appropriate interactive tools and graphical visualization, to highlight the properties of the challenging scenarios. In response to these observed properties, policy devices can be crafted to meet the challenges.

With a newly enlarged set of possible strategies, the search can continue for those that do well across wide ranges of futures and for futures that break candidate strategies. As surmised, there exist adaptive milestone strategies more robust than any static strategy. More aggressive computer searching reveals futures in which these new candidates also fail. Consequently, the process again follows arc "d." This results, in the fifth section, in building another adaptive layer on top of milestones, which results in a new policy candidate, "Safety Valve." This strategy appears to be robust in nearly all futures, but additional search still discovers some exceptional cases that break it.

At this point, one could follow the arc one more time and build adaptive mechanism into this safety valve strategy to deal with the troublesome cases. But analyses of long-term policy will often reach a point where, due to a shortage of time, resources, or creativity, a conclusion must be reached without further experimentation. So, a decision is made at this point to consider the question of whether, in spite of the remaining troubling futures, society should choose to adopt the Safety Valve strategy. Thus, the sixth section finally invokes probabilistic weighting, but not by guessing what the correct probabilities should be. Instead, the probabilities associated with the troublesome futures may be varied to determine what one would have to believe about the relative likelihood of such futures to bet that the Safety Valve strategy has a lower expected outcome than does a competitor. As described in the seventh section, this approach provides a powerful and systematic means to characterize deeply uncertain, heretofore "unquantifiable" surprises. The analysis takes a complex problem with a huge number of dimensions and reduces it to a small number of crucial bets that can be passed along to senior decisionmakers who must make the final judgment call.

In the absence of predictability, no conclusion is ever final. However, the human-machine collaborative search for robust policy options, interleaved with human-machine collaborative search for plausible scenarios that will break a given candidate strategy, can be used to ensure that available resources are brought to bear to discover the best policy option possible.

LANDSCAPES OF PLAUSIBLE FUTURES

At each stage of the robust decisionmaking process, interactive computer visualizations provide a powerful tool to help humans explore patterns and other properties of the large, multidimensional data sets created during this disciplined search for robust decision strategies. Figure 5.1 shows one such "landscape" of plausible futures. The landscape's horizontal and vertical axes depict two key uncertainties that the sustainability debate suggests are particularly important in distinguishing among alternative futures: the average global rates for economic growth and for decoupling—that is, the reduction in emissions intensity per unit of economic output during the twenty-first century. Each point of intersection of the possible values for the two variables defining this landscape represents a particular scenario, in this case defined narrowly by holding all other variables constant.[1] The upper left-hand region portrays futures where decoupling reduces pollution much faster than the economy grows. The lower right-hand region portrays futures where the

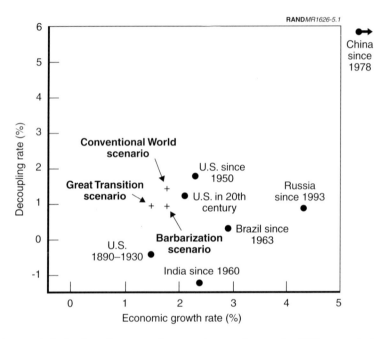

NOTE: Labeled points show actual economic growth rates (in GDP per capita) and decoupling rates (for CO_2) of the United States, China, India, Brazil, and Russia over various periods. Plus signs show the economic growth and decoupling rates used in Wonderland recreations of the GSG Conventional World, Barbarization, and Great Transition scenarios as shown in Figures 5.2 and 5.3. Assumptions about other key driving forces for these scenarios not shown on this landscape are given in Table B.1.

Figure 5.1—Landscape of Plausible Futures as a Function of Global Economic Growth Rates and Decoupling Rates over the Twenty-First Century

---

[1]The views are static representations generated using Computer Assisted Reasoning® system (CARs™) software. CARs operates dynamically so that, in using the system, those variables not actively represented in any view are present as slider bars in the graphic user interface. These may be moved individually or in groups to see how changes in other variables would affect the landscape views on the screen.

economy grows much faster than decoupling reduces pollution. Inwhat follows, color-coded patterns will be employed to compare the performance of alternative strategies across this landscape. Here, the figure shows only a blank landscape in order to describe how it portrays the relationships among scenarios.

The landscape aims to capture the full range of plausible scenarios for the twenty-first century. It depicts average economic growth rates ranging from a catastrophic century-long deflationary trend to an unprecedented level of sustained high global growth. The span of decoupling rates is similarly heroic. They range from negative rates (indicating economies growing increasingly pollution-intensive over time) characteristic of history's most environmentally unfriendly periods of industrialization to unprecedented rapid diffusion of environmentally friendly technology and practices. Anchor points (black dots) on the landscape serve to compare its scenarios to the range of historical trends. This landscape spans a much wider range than characterized the changes in the United States over the nineteenth and twentieth centuries but a narrower range than that spanned by the performance of several other nations over the last several decades.[2] The landscape portrays global, century-long futures, whose plausible range should be less than that experienced over

---

[2]For the purposes of this illustrative exercise, the anchor points on this figure only show the data for CO_2 emissions. As the United State industrialized between 1890 and 1930, its GDP per capita grew annually at about 1.5 percent on average while the amount of pollution per economic activity increased annually by 0.45 percent (a negative decoupling rate of –0.45 percent). Over the entire twentieth century, both economic growth and decoupling were higher, and higher still during the half century since 1950. Even as economic growth has increased, the decoupling rate has also improved. This acceleration in both economic growth and decoupling that reduces its adverse consequences is a prime driver of the optimists' confidence that sustainability is not a problem. During the past few decades, a number of developing countries have grown much faster than the United States, but they have had a wide range of decoupling rates. China has been high in both, in large part because it has retired a legacy of inefficient Communist-era plants. In contrast, India's per capita economic growth since 1960 has come with negative decoupling rates. U.S. GDP data for 1929 to the present are from the U.S. Bureau of Economic Analysis (BEA); 1869–1929 data are from Romer (1989); U.S. population from 1929 to the present is from the U.S. BEA; 1880–1928 population data are from the U.S. Department of Commerce's Historical Statistics of the United States; economic and population data for China, India, and Brazil are from the International Monetary Fund website (http://imfStatistics.org); data for Russia are from the European Bank's Transition Report Updates (April 2001 and 2002); all CO_2 emissions data are from Marland et al. (2002).

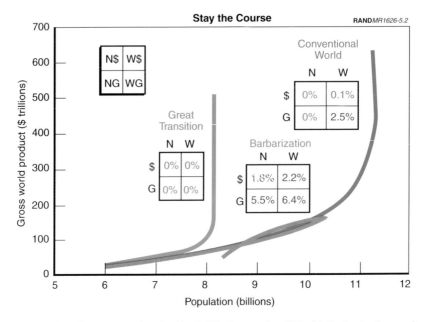

NOTE: These futures were inspired by GSG's Conventional World, Barbarization, and Great Transition scenarios. Boxes show regret of the Stay the Course strategy in each future for each of four measures of the future human condition: North Quasi-HDI (N$), World Quasi-HDI (W$), North Green HDI (NG), and World Green HDI (WG). See Table B.1 for values of Wonderland input parameters used to create these scenarios.

Figure 5.2—Comparative Trajectories of Output per Capita and Population for Three Futures for the Stay the Course Strategy

smaller regions and shorter periods. Thus these historical anchor points suggest that, at least as an initial assumption, the landscape is sufficiently expansive to capture the plausible range of future scenarios for the twenty-first century.

Of course, many factors other than global economic growth and decoupling rates may prove decisive over the twenty-first century. It is thus important to note that the landscape shown in Figure 5.1 actually represents a two-dimensional slice through a multidimensional space whose axes are defined by each of the uncertainties in the analysis. This makes an important point: Rather than assume away uncertainties by either dropping poorly understood factors from consideration or assigning them arbitrary values, they are retained and explored over their full plausible range. From among

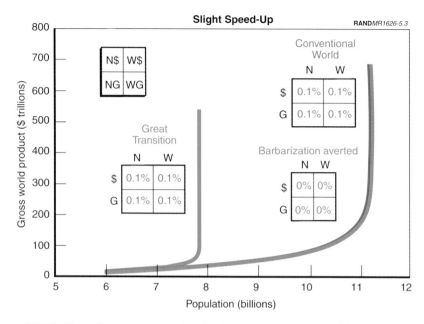

NOTE: These futures were inspired by GSG's Conventional World, Barbarization, and Great Transition scenarios. Boxes show regret of the Slight Speed-Up
strategy in each future for each of four measures of the future human condition:
North Quasi-HDI (N$), World Quasi-HDI (W$), North Green HDI (NG), and
World Green HDI (WG). See Table B.1 for values of Wonderland input parameters used to create these scenarios.

**Figure 5.3—Comparative Trajectories of Output per Capita and Population
for Three Futures for the Slight Speed-Up Strategy**

the very large number of potential low-dimensional landscapes one
could create, successful tools for LTPA must help users identify a
small set that reveals important patterns and then discover any
caveats or surprises caused by parameter combinations not shown in
these particular views. For instance, during the course of this exercise, we will confirm the importance of the two uncertainties
explored in Figure 5.1, using human-guided, interactive explorations
across many alternative landscapes, computer-generated searches
for special cases, and formal sensitivity analysis that gives statistical
summaries of the properties of the scenario ensemble.

A key purpose of such landscapes of plausible futures is to clarify the
relationship among different scenarios. For instance, in Figure 5.2

the modified "Wonderland" scenario generator reproduces three paths into the future corresponding to one representative scenario from each of GSG's Conventional Worlds, Barbarization, and Great Transitions families. These three futures may be represented by choosing appropriate input parameters for average economic growth rates and decoupling rates, relative rates of economic growth in the North and South, seriousness of future environmental constraints, and criteria/values used by future decisionmakers in responding to the sustainability challenges they face. Comparison of Figure 5.2 with Figure 2.1 demonstrates that the Wonderland-generated paths have somewhat higher economic growth rates but otherwise roughly correspond to those of GSG.

These points may be laid on the landscape to understand their relationship to each other. The points marked with plus signs on Figure 5.1 show that the generated Conventional Worlds and Barbarization scenarios have the same rate of exogenous growth, 1.8 percent, and differ in their innovation rate by only 0.5 percent (1.5 percent and 1 percent for the two scenarios, respectively). Barbarization and Great Transitions share the same decoupling rate, but the latter has a slightly lower growth rate. These three divergent futures are, in fact, rather closely grouped. The landscape of plausible futures suggests that the three GSG scenarios represent only a narrow slice of the potential triumphs and pitfalls of the twenty-first century.

NO FIXED STRATEGY IS ROBUST

The pieces are now in place for a first attempt to identify robust near-term strategies that can ensure long-term sustainable growth by proceeding along the "a" path in the procedure illustrated in Figure 3.1. Creation of the scenario ensemble uses the range of futures shown in Figure 5.1. The initial strategies consider fixed near-term strategies of the type shown in Figure 4.1 in which policymakers pursue specific near-term actions, such as R&D and environmental taxes, that increase the decoupling rate by some prescribed amount.

Figure 5.4 depicts the performance of one such strategy, Slight Speed-Up, across the landscape of plausible futures. Slight Speed-Up increases by 1 percentage point the rate of decoupling in both North and South over the respective rates that would otherwise hold

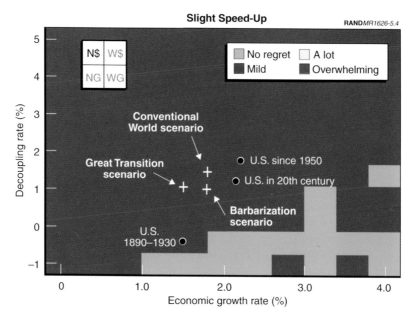

NOTE: Green areas have no regret; blue areas have relative regret of less than 0.1 percent; yellow areas have relative regret less than 2 percent; red areas have relative regret greater than 2 percent. The small four-quadrant box indicates the measure of human condition used in each panel: N$ for North Quasi-HDI, NG for North Green, W$ for World Quasi-HDI, and WG for World Green. Dots show historical data for U.S. economic growth rates (in GDP per capita) and decoupling rates (for CO_2) for the United States over various periods. Plus signs show the economic growth and decoupling rates used in our Conventional World, Barbarization, and Great Transition scenarios (these scenarios also differ in the value of factors not shown in this landscape).

Figure 5.4—Performance of Slight Speed-Up Near-Term Strategy over a Landscape of Plausible Futures Using the North Quasi-HDI Measure

in the absence of any policy intervention. The figure shows the regret of Slight Speed-Up compared to the 15 other alternative strategies shown in Table B.2, across the range of plausible futures using the North Quasi-HDI measure of the future human condition. For each future depicted in this landscape, the computer uses the scenario generator to calculate the performance of Slight Speed-Up.[3]

----

[3]Having selected a particular granularity for dividing the full span of values on each axis, the landscape presents the results for Slight Speed-Up across 121 alternative futures.

It then finds the fixed strategy that produces the highest (optimal) performance in that future using the North Quasi-HDI measure. Slight Speed-Up's regret in any future is defined as the difference between its performance and the optimum (see Equation 3.1). The landscape uses colors to indicate patterns in Slight Speed-Up's regret.

Slight Speed-Up has mild regret throughout almost the entire landscape and has no regret in several futures where economic growth rates greatly exceed the decoupling rate. While this landscape shows scenarios with a wide variation of assumptions about future global average economic growth and decoupling rates, the uncertainties in a wide range of other factors remain unexplored. For instance, all the futures in the landscape use a single value for the costs of implementing Slight Speed-Up, calibrated to the estimate that the current level of environmental protection in the developed world has cost about 2 percent of GDP. Nonetheless, this first view across the scenario ensemble suggests that Slight Speed-Up might be robust.

However, Figure 5.5 forces a change in our assessment. This landscape is identical to the first except that it assesses Slight Speed-Up's performance in each future using the World Green-HDI measure. From this vantage point, the strategy has mild regret over those regions where the decoupling rate exceeds economic growth. Its regret is overwhelming in futures where economic growth rates exceed decoupling rates. In fact, viewed from the perspective of the World Green-HDI measure, Slight Speed-Up offers few advantages over the landscape than does the Stay the Course strategy, shown in Figure 5.6, which takes no near-term actions to address sustainability. As shown in Figure 5.6, Stay the Course depicts no regret over those regions where the decoupling rate exceeds economic growth because, in such futures, no policy response is necessary. Resembling Slight Speed-Up, Stay the Course has overwhelming regret in futures in which economic growth rates exceed decoupling rates. Only in a narrow band of futures would Slight Speed-Up offer better outcomes than Stay the Course. Slight Speed-Up no longer appears to be so robust.

In fact, no fixed near-term strategy of the form shown in Figure 4.1 appears robust over all four measures as well as over the full range of

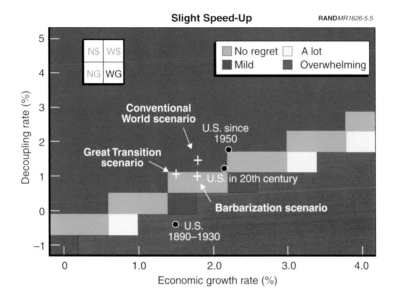

Figure 5.5—Performance of Slight Speed-Up Near-Term Strategy over a
Landscape of Plausible Futures Using World Green Measure

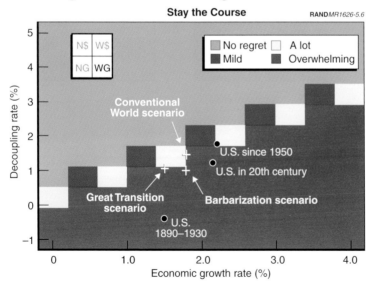

Figure 5.6—Performance of Stay the Course Near-Term Strategy over a
Landscape of Plausible Futures Using World Green Measure

plausible futures described by the variables discussed in the context of XLRM in Chapter Four. Figure 5.7 shows the performance of a Crash Effort strategy, which increases decoupling in both North and South by 2 and 4 percentage points, respectively, above their exogenous values over the landscape for both the North Quasi-HDI and World Green-HDI measures. This strategy has mild or no regret in those futures where Slight Speed-Up has overwhelming regret. However, Crash Effort has a lot of regret in those futures where Slight Speed-Up performs well. Clearly, Slight Speed-Up offers an insufficient response in futures that most demand near-term action, while Crash Effort represents an overreaction in the futures where the sustainability problem will largely solve itself.

RAND*MR1626-5.7*

Crash Effort

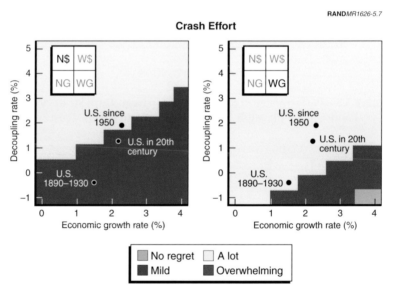

NOTE: Regret was calculated in comparison to 15 other alternative innovative strategies shown in Table B.2. Green areas have no regret; blue areas have relative regret of less than 0.1 percent; yellow areas have relative regret of less than 2 percent; red areas have relative regret greater than 2 percent. Anchor points show historical U.S. economic growth rates and decoupling rates.

Figure 5.7—Performance of Crash Effort Near-Term Strategy over a Landscape of Plausible Futures Using Quasi-HDI and World Green HDI Measures

Implementing Robust Decisionmaking 101

This dilemma appears inescapable for any fixed near-term strategy. We examined 16 strategies representing a wide range of different combinations of decoupling rate increases in North and South. (These 16 were all combinations of policies that increase decoupling in North and South by 4, 2, 1, or 0 percentage points. Three of the 16 have names: Crash Effort [2 percent, 4 percent], Slight Speed-Up [1 percent, 1 percent], and Stay the Course [0 percent, 0 percent].) A computer search was constructed for each strategy using each of the four measures to look for regions of large regret over the full range of plausible futures.[4] Each of the 16 innovation strategies fails overwhelmingly over a wide range of futures using one or more measures. This failure is no surprise because these strategies poorly exploit the third element of LTPA: They are not sufficiently adaptive.

Given this result, it is interesting to compare the fate of the scenarios inspired by those from the three GSG families under the Slight Speed-Up and Stay the Course strategies. The paths shown in Figure 5.3 assume Stay the Course. The four quadrants by each path show this strategy's regret with respect to each of the four measures of the future human condition. The labeled points in Figure 5.6 show the assumptions about the economic and decoupling rates used to create the paths in Figure 2.1. In Barbarization, Stay the Course has considerable regret when the results are assessed using the North Quasi-HDI measure and overwhelming regret viewed with the other three measures. In Conventional Worlds, the strategy has no regret viewed with the North Quasi-HDI and World Quasi-HDI measures, mild regret viewed with the North Green-HDI, and overwhelming regret viewed with World Green-HDI measure. This last result, which is also seen in Figure 5.6, occurs because in this scenario future decisionmakers avail themselves of an option to adapt to their rapidly degrading environment, thus preserving economic growth but not environmental quality. Only in Great Transitions does Stay the Course prove uniformly successful. The strategy has no regret when viewed with all four measures of the future human condition. Great Transitions assumes near-term changes in values that alter

---

[4]The simple global search routine launched "downhill" searches at each of a hundred randomly chosen starting points in the space. While this algorithm is not necessarily an efficient means of finding a global maximum, it is sufficient to generate breaking scenarios to demonstrate that the fixed strategies are not robust.

economic growth paths and make additional increases in decoupling rates unnecessary.

In contrast, Figure 5.3 shows the path produced by the same futures as shown in Figure 5.2 but with a choice of the near-term Slight Speed-Up strategy. The results are dramatically different. In particular, the strategy has rendered Barbarization a no longer unpleasant future. It now follows virtually the same path as Conventional Worlds. If the future were guaranteed to follow either the Conventional World, Barbarization, or Great Transitions futures, Slight Speed-Up could lay strong claim as a robust strategy. It performs well, yielding no or mild regret, in all three futures using all four measures of the future human condition.

GSG proposed its three scenarios to support a risk management argument leading to the conclusion that a near-term effort to promote a Great Transitions future is necessary to avoid the risks of falling into Barbarization. However, Slight Speed-Up's ability to divert the Barbarization future along a more desired path while performing well in the Conventional World and Great Transition futures undercuts this claim.[5] As suggested by Figures 5.4 and 5.5, Slight Speed-Up might eliminate the need for Great Transitions if policymakers were certain that Conventional Worlds and Barbarization were the only other futures that might lie ahead. GSG would quite rightly object that their families of scenarios contain much greater richness and detail than the three Wonderland-generated paths shown here. They would point to other potential weaknesses of Barbarization not addressed by Slight Speed-Up. This is precisely the point. Given the multiplicity of plausible futures, any policy argument based on a small number of scenarios or families can be easily challenged by questioning assumptions and pointing to paths not included in the analysis.

Proponents on many sides of the sustainability debate make risk-based arguments. Environmentalists, such as the GSG, point to the

---

[5]The GSG's simulation models include a far greater range of factors than the simple scenario generators used here. It is unlikely that small changes in any single input parameter would cause one of their scenarios to follow a significantly different course. However, small changes in a number of key assumptions could well make one of their model-generated scenarios tell a very different story.

environmental and social risks of society's current development path. Many international businesses and governments of developing countries emphasize the risks posed by sustainability policies that might hamper economic growth. A robust decision analysis full-scenario ensemble provides a much firmer foundation to assess such arguments than do any small set of scenarios. While Slight Speed-Up is robust over GSG's Conventional Worlds, Babarization, and Great Transitions, it can easily fail catastrophically for a variant of Barbarization described by a slightly different set of assumptions. This is clearly seen by the proximity of Barbarization to the regions of overwhelming regret in Figure 5.5. Such landscapes thus make explicit the concerns policymakers naturally have with any LTPA based on a small number of scenarios. With a restricted sample, it is never clear how the conclusions might change if some seemingly innocuous assumptions, whether made by industrialists or environmentalists, were subtly changed. A strategy found robust across the full landscape of plausible futures will be less vulnerable to such criticisms.

EXPLORING NEAR-TERM MILESTONES

In the analysis so far, no fixed near-term strategy appears robust across the full range of plausible futures and values. Moderate near-term actions fail in futures requiring drastic interventions. Aggressive near-term strategies impose unwarranted costs in futures where sustainability is no problem. No fixed strategy appears to guarantee the appropriate level of policy-induced decoupling across the entire landscape of plausible futures. We thus begin another pass through the procedure shown in Figure 3.1. Based on what we have learned, we augment our menu of alternative strategies with the aim of crafting ones that may prove more robust.

Policymakers have a well-established, often successful approach when deep uncertainty prevents them from prescribing in detail the series of steps needed to reach well-understood goals. In such circumstances, they often set performance milestones to be achieved at some point. They then implement an adaptive strategy, adjusting actions over time in an attempt to achieve the desired ends. This, for instance, is the approach the UN has attempted in setting its Millennium Development Goals.

Adaptive strategies may be implemented by incorporating near-term milestones within the modified "Wonderland" scenario generator as shown in Figure 5.8. Policymakers set a near-term goal for emissions intensity in the North and South—that is, the difference between the economic growth rate and the policy induced decoupling rate. Society then takes whatever steps are needed to achieve these milestones by the appointed time. These policies remain in place until future decisionmakers detect some change in the environmental carrying capacity and implement a new set of policies. For instance, the No Increase strategy demands that policymakers choose annual interventions that hold emissions intensity (emissions per unit of GDP) constant in both North and South after 2010. The diagram in Figure 5.8 suggests that analysts can represent such strategies by making appropriate changes in the computer code for the scenario generator (a step indicated by line "d" in Figure 3.1). The equations and numerical values are described in Appendix B.

We can now begin to explore the robustness of these innovation strategies using landscapes similar to those in Figures 5.4, 5.5, 5.6,

RAND*MR1626-5.8*

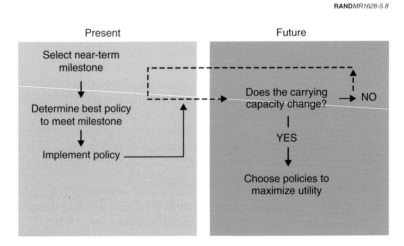

NOTE: Numerical values and equations describing the strategies can be found in Table B.2.

Figure 5.8—Milestone Strategy

and 5.7. Figure 5.9 depicts the performance of the No Increase strategy, compared to the 15 alternative milestone strategies that set the maximum increase in the difference in economic growth rate and policy-induced decoupling rate in North and South to some combination of 0 percent, 1 percent, 2 percent, or No Limit, where No Increase is the (0 percent, 0 percent) case. The regret of No Increase is shown over the landscapes of plausible futures using the North Quasi-HDI and World Green-HDI measures as a function of the exogenous economic growth and decoupling (that is, before any policy intervention) rates. In futures where exogenous decoupling outperforms economic growth, no interventions are needed to meet

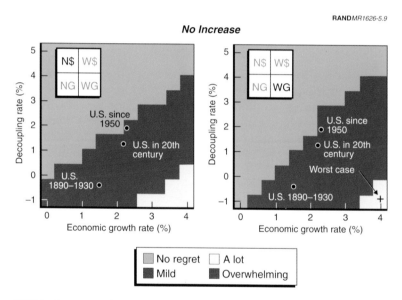

RAND*MR1626-5.9*

No Increase

NOTE: Green areas have no regret; blue areas have relative regret of less than 0.1 percent; yellow areas have relative regret of less than 2 percent; red areas have relative regret greater than 2 percent. A plus sign marks the economic growth and decoupling rates which yield No Increase's worst-case future, though this future also differs in the values for the factors not shown on this landscape. Regret is calculated as a comparison to 15 alternative milestone strategies shown in Table B.3.

Figure 5.9—Performance of No Increase Near-Term Milestone Strategy over a Landscape of Plausible Futures Using North Quasi-HDI and World Green Measures

the milestone and the strategy therefore imposes no costs. In futures where decoupling lags, the strategy calls for policies that increase decoupling. In all but the most severe cases (the lower right-hand corner), these policies impose small near-term costs but prove suffi- cient to preserve sustainability over the long term.[6]

At first glance, No Increase appears robust. The analytic process thus shifts to testing this hypothesis. Policymakers who have risen to the top usually also have a good working understanding of Murphy's Law and frequently ask, "What can go wrong?" The computer can help address this question by finding and assessing plausible futures where No Increase does poorly—that is, where it has high regret. In particular, a computer search can identify sets of input parameters to the scenario generator that yield high regret, and sensitivity analysis can suggest which of those key inputs—i.e., the key driving forces— are most responsible for the strategy's poor performance.[7]

The landscape in Figure 5.10 suggests how such information can help answer the policymakers' critical question. The figure depicts the performance of No Increase, viewed with the World Green-HDI measure across a landscape that includes the strategy's worst-case future. A simple computer search (represented by line "b" in Figure 3.1) identified this worst case.[8] The landscape's horizontal and verti-

---

[6]This report does not compare No Increase to other potential milestone strategies. The discussion in Chapter Six suggests that other strategies perform better than No Increase. This does not change the conclusions of any of the discussions here.

[7]Global sensitivity analysis uses statistical methods on a very large sample of points over the multidimensional landscape to identify those input factors whose variation is most important in producing variation in the output parameter of interest, in this case the strategy's regret. See discussion in Chapter Six.

[8]A simple "downhill" search was launched at each of 500 randomly chosen starting points in the space of input parameters representing the exogenous uncertainties (X parameters of Table 4.1) in the Wonderland scenario generator. The search aimed to maximize the regret of the No Increase strategy. This process yielded 1,502 scenarios because we retained in our scenario ensemble all the intermediate steps produced by the search algorithm. Only those scenarios were kept whose demographic input parameters produced plausible future population scenarios, defined as a 2050 popu- lation between 700 million and 1.5 billion in the North and a 2050 population between 7 billion and 15 billion in the South. The 837 cases that did not meet this criteria were discarded. Of the remaining 665 cases, the one with the largest No Increase regret was selected as the worst case.

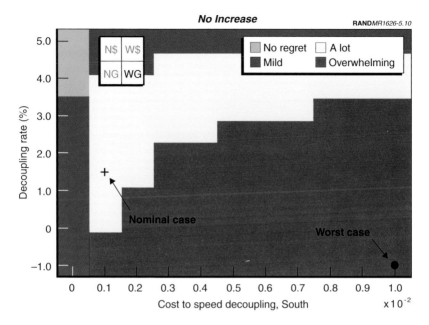

NOTE: Landscape includes worst-case future for No Increase found by computer search. Assumed values for other uncertainties are shown in Table B.4. A plus sign marks the costs and decoupling rates used in the landscapes in Figures 5.4, 5.5, 5.6, 5.7, and 5.9 although the values for other factors differ from those shown in this landscape.

Figure 5.10—Performance of No Increase Strategy over a Landscape of Plausible Futures, Including No Increase's Worst-Case Future

cal axes depict two key uncertainties that the computer's sensitivity analysis suggests are critical to the strategy's performance[9]: cost of speeding decoupling in the South and exogenous decoupling rate (that is, before application of any policy intervention).[10] The nominal case for the former uncertainty, used in landscapes of Figures 5.4, 5.5, 5.6, 5.7, and 5.9, corresponds to assumptions that current levels

---

[9]We discuss these sensitivity analysis results in Chapter Six.

[10]In theory, the analyst may search for such modes manually by changing the axes and types of views requested by the computer and moving the slider bars through the full range of uncertainty for each variable. As a practical matter, this would be subject to possible unintentional bias or merely fatigue. Hence, the value in automating such searches for failure modes.

of environmental quality have been purchased at a cost of about 2 percent of GDP and that future policies in the North and South will show similar costs. In the worst case for No Increase, the future costs of increasing decoupling in the South are about 10 times higher than the historic values. The decoupling rate in the Conventional World scenario is 1.8 percent. This landscape suggests that, with this level of exogenous decoupling, No Increase can fail (have overwhelming regret) in cases where the costs of increasing future decoupling have any value greater than about three times historic levels.

Yet in this landscape No Increase performs poorly even with the nominal assumptions about cost and decoupling (indicated by the point marked "Nominal Case") as opposed to the strategy's good performance in the nominal case in the landscape of Figure 5.9. This variation results from the different values for the parameters not shown in the two landscapes. In particular, the computer search suggests that the No Increase worst case also occurs with very high economic growth rates of 4 percent in the North and 7.5 percent in the South and few environmental constraints as represented by a level of sustainable pollution nearly 10 times current levels. In addition, it is also blessed with future decisionmakers who have little inclination or capability to respond to sustainability threats as represented by high discount rates and poor ability to detect changes in carrying capacity. (Table B.4 provides a full listing of parameter values for the No Increase worst case.)

Drilling down at selected points on the landscape yields insight into the factors that shape the No Increase patterns of regret. In particular, Figure 5.11 compares the trajectories for two types of output per capita: those in a twenty-first century North and South that result from a near-term choice of the No Increase strategy in its worst-case future and those that result from a near-term choice of the optimal strategy in that future. The best alternative to No Increase in its worst case is M0X, which sets a stringent milestone in the North (no difference between decoupling rate and economic growth rate) and no milestone in the South. The strategy successfully exploits this future's low sustainability threat (represented by high sustainable pollution thresholds) and the low costs to speed decoupling in the North to yield a World Green-HDI performance of 6, a truly significant improvement of the human condition. In contrast, the stringent

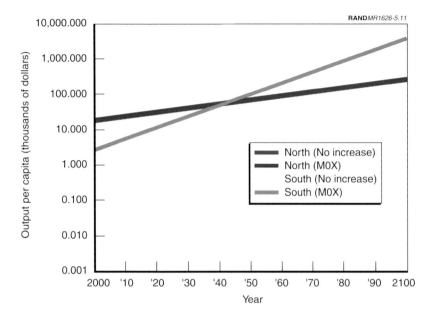

NOTE: The trajectories will result from the strategies in the worst-case future depicted in Figure 5.10.

Figure 5.11—Comparative Trajectories of Output per Capita for the Twenty-First Century for the No Increase and M0X Strategies in the No Increase Worst-Case Future

Southern milestone imposed by No Increase collapses economic growth in the South because of the extremely high costs of speeding decoupling in this region. Viewed in context of the World Green HDI, No Increase results in a dismal performance of –7 percent. Compared to the possibilities this future offers, the results from No Increase yield an enormous regret of 13 percent.

Once the computer has revealed this result, the analysts, experts, and policymakers can address the issue of whether such poorly performing futures are sufficiently plausible to cast doubts on the efficacy of the No Increase strategy. There is certainly much to question about No Increase's worst case. Economic growth rates, the resilience of the environment, and the costs of reducing pollution are exceedingly high compared to today's norms. The scenario assumes policymakers would cling to the Southern decoupling milestone in the face of a catastrophic drop in economic output. It is worth noting that one of this future's most salient features is that, by century's end, the South will be as much richer relative to the North as the North is richer relative to the South today. This is not a prognostication likely to be included in the result from any group of experts generating narrative scenarios in today's world of American hegemony.

Nonetheless, the basic outlines of this scenario are precisely those articulated by the camp in the sustainability debate that argues that too-aggressive policies to protect the environment will prevent otherwise attainable and spectacular improvements in the human condition. The vast sea of red in Figure 5.10 suggests that one need not fully embrace the potentially extreme assumptions underlying the No Increase worst case to entertain doubts about the strategy's robustness. Computer-guided explorations of the scenario ensemble could test this judgment by allowing stakeholders to modify key assumptions about growth rates, costs, environmental resilience, and the behavior of future decisionmakers and determine the extent to which No Increase still fails among a more restrictive set of cases. It is not hard to show that No Increase and all the other alternative milestone strategies can fail in futures to which many stakeholders subscribe and that cannot now be disproven by conclusive evidence.

IDENTIFYING A ROBUST STRATEGY

Ultimately, judgments about the robustness of strategies depend on the available alternatives. The adaptive Milestone strategies perform generally better than nonadaptive Innovation strategies but are still not robust. In particular, No Increase fails when the cost of reaching its near-term goals proves too high. Once again, we pass along path "d" in Figure 3.1, attempting to design new strategies that may be more robust than the current alternatives.

Policymakers often set contingent goals. For instance, an agreement to buy a house may be subject to the property's passing an inspection. In the climate-change arena, some have proposed contingent targets as a means to address the deep uncertainties over the costs of meeting the near-term emissions reduction milestones of the Kyoto Protocol. Such a Safety Valve strategy (Victor, 2001) might be implemented, for instance, with a tradeable pollution system in which the government agrees to sell an unlimited number of permits at some maximum price.

Such adaptive, contingent strategies may be represented within the modified "Wonderland" scenario generator as shown in Figure 5.12. Policymakers set a near-term goal for emissions intensity in the North and South—that is, the difference between economic growth and policy-induced decoupling rates. However, the annual costs are monitored and, if they ever exceed some predetermined threshold, the policies are relaxed to meet the cost target, irrespective of whether or not society will achieve the level of decoupling improvement required to meet the emissions intensity milestone. The Safety Valve strategy used here aims for no increase in emissions intensity by 2010 in both North and South, achieved at a cost of no more than 1 percent of GDP in each region.

Figure 5.13 depicts Safety Valve's performance over the landscape of plausible futures compared to the 16 alternative milestone strategies

RAND*MR1626-5.12*

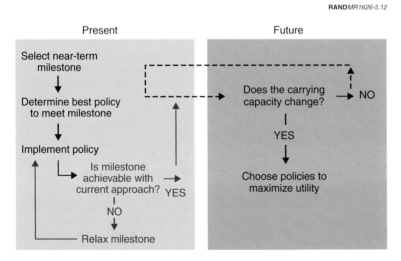

NOTE: Numerical values and equations describing the strategies can be found in Appendix B.

Figure 5.12—Safety Valve Strategy

shown in Table B.5. Across these futures, its performance is similar to that of No Increase except in the high-growth, low-decoupling corner, where it performs much better. In these futures, the level of policy intervention needed to achieve No Increase's milestones proves too costly relative to Safety Valve's more measured response.

Is Safety Valve robust? Once again, the computer can help address the policymaker's key question, "What can go wrong?" A computer search reveals that Safety Valve can still fail, but the search also provides important information that can help human decisionmakers characterize and assess these potential failures and thus make informed judgments as to the robustness of the Safety Valve strategy and its potential alternatives.

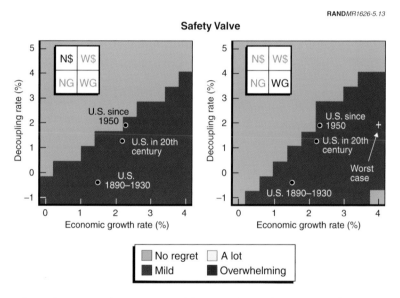

NOTE: Green areas have no regret; blue areas have relative regret of less than 0.1 percent; yellow areas have relative regret of less than 2 percent; red areas have relative regret greater than 2 percent. A plus sign marks the economic growth and decoupling rates used in Safety Valve's worst-case future, though this future differs in the values for the factors not shown on this landscape.

Figure 5.13—Performance of Safety Valve Near-Term Strategy over a Landscape of Plausible Futures Using North Quasi-HDI and World Green Measures

Figure 5.14 shows the performance of Safety Valve as a function of
two key parameters identified by a sensitivity analysis as most critical
in producing large regret for this strategy: the importance future
Southern decisionmakers place on the environment as represented
by the weightings in their utility function (horizontal axis) and the
sustainable level of pollution in the South (vertical axis). The input
parameters not shown by the axis of this landscape are, as before,
those defined by the worst-case regret found by the search as listed
in Appendix B. The slice of the landscape shown in this figure is
characterized by high economic growth rates in the North (3.9 percent) and South (6.2 percent), relatively slow decoupling rates in the
North (1.9 percent) and South (–1 percent), future decisionmakers

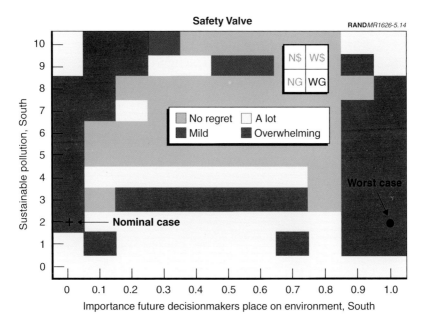

NOTE: Landscape includes worst-case future for Safety Valve found by computer
search. Assumed values for other uncertainties are shown in Table B.4. A plus sign
marks the parameter values used in the landscapes in Figures 5.4, 5.5, 5.6, 5.7, 5.9,
and 5.13 although the values for other factors differ from those shown in this
landscape.

**Figure 5.14—Performance of Safety Valve Near-Term Strategy over a
Landscape of Plausible Futures, Including Safety Valve's Worst Case**

who are not particularly effective at responding to sustainability challenges (modeled by high discount rates and relatively low ability to detect changes in natural capital), and high costs of implementing near-term innovation policies (five and eight times the nominal case in North and South, respectively).

The "nominal" values for the level of sustainable pollution and the importance future decisionmakers place on the environment are shown in the lower left-hand corner. These are the values used in the landscapes in Figures 5.4, 5.6, 5.7, 5.9, and 5.13. The worst case for the Safety Valve strategy, shown in the lower right-hand side of the figure, has roughly the same level of sustainability challenge (a sustainable level two times current emissions) but a much higher importance placed by future decisionmakers on the environment. The strategy's performance is highly nonlinear over this landscape.

Figures 5.15 and 5.16 suggest the reasons for Safety Valve's poor performance in its worst-case future. The figure compares the trajecto-

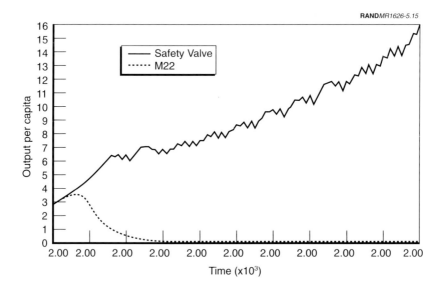

Figure 5.15—Trajectories of Output per Capita in the South for the Safety Valve and M22 Strategy in the Safety Valve's Worst-Case Future

ries for output per capita in the South (Figure 5.15) and for Southern death rates (Figure 5.16) for Safety Valve in its worst case and for the optimum strategy in that future. The optimum strategy is M22, which sets relatively lax, but nonetheless inviolate, milestones for both North and South. Given the high growth rates and high costs of speeding decoupling, the M22 strategy devastates the Southern economy. In contrast, Safety Valve makes only mild attempts to speed decoupling and thus preserves the economic expansion. Nonetheless, Safety Valve does much worse in advancing the human condition because, in stark contrast to No Increase's worst case, this future contains significant environmental constraints. In particular, the human population proves extremely sensitive to environmental degradation. Thus, in allowing the pollution to grow unabated, Safety Valve precipitates waves of epidemics that devastate the population. In this unenviable future, M22 yields a dismal World

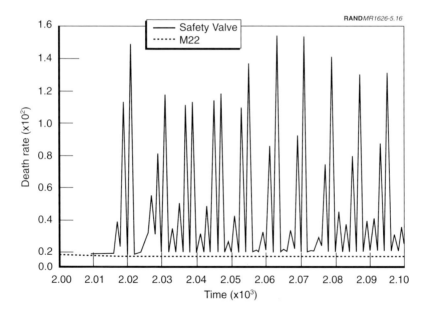

Figure 5.16—Trajectories of Death Rates in the South for the Safety Valve and M22 Strategy in the Safety Valve's Worst-Case Future

Green HDI of –2 percent. But Safety Valve yields a horrible –10 percent for a regret of 8 percent.

The Safety Valve worst case is in many ways a perverse scenario. Future Southern decisionmakers care nothing about human health (this term is not included in their utility function). The frequent population collapses reduce the impact of economic growth on the environment, so these decisionmakers take no action to slow the increase in pollution. M22 succeeds because it reduces pollution sufficiently and carrying capacity never drops, the second decision period of Figures 5.8 and 5.12 never begins, and thus the future decisionmakers are never placed in a position where they can make these decisions that we find so perverse. In an analysis meant to inform actual policy judgments, this scenario would likely provoke two responses. Participants might reject it out of hand, remove it from the scenario ensemble, and search for other, more credible futures where Safety Valve fails. Participants might also return to the scenario generator code and modify its representation of future decisionmakers to eliminate their perverse behavior.

Nonetheless, this worst-case scenario also raises two important methodological points that are our main interest here. First, the future shown in Figures 5.15 and 5.16 differs in at least one profound way from that shown in Figure 5.11: It offers policymakers no good options. The Safety Valve worst case presents a future when even the optimal combination of policy levers considered in Table 4.1 produces a dismal outcome. Faced with such a situation, participants in an actual policy-relevant exercise might follow path "d" in Figure 3.1 and seek other policy options that might prove effective in such futures. A variety of levers would be potentially useful in Safety Valve's worst case. Public health and education policies might reduce the vulnerability of human population to declines in environmental carrying capacity and, thus, the high death rates shown in Figure 5.16. R&D funding and institutional reform might lower the cost of speeding the decoupling rate. Near-term policies that promoted improved environmental monitoring, shaped people's future values, and improved the efficiency of governing institutions might increase the ability of future policymakers to detect and respond to imminent collapse of carrying capacity. The simple Wonderland scenario generator used in this study will not, however, support consideration of any such policies. Modifications that would enable it to do so lie well beyond the scope of this study. An improved scenario generator would, however, allow analysts to continue iterating through the steps of Figure 3.1, design actions effective against such futures as the Safety Valve worst case, assess their performance across a wide range of plausible futures, and eventually find strategies that are even more robust.

No matter how sophisticated the scenario generators become, any long-term policy analysis in a sufficiently challenging policy environment may ultimately reveal futures against which no strategy is entirely robust. This situation may be an unavoidable characteristic of the world itself rather than any computer code used to describe it. In some futures, no combination of available policy levers may offer a good outcome. More vexingly, some stressing futures may demand a strategy that has unavoidably high regret in a wide range of other cases. For instance, successfully addressing the Safety Valve worst case might demand expensive measures to reduce costs and to adapt to environmental changes unneeded in most other futures. Thus, it might be tempting to ignore such potentially perverse scenarios. Before discarding them entirely, however, it is worth remembering that perverse scenarios do occasionally unfold. The Safety Valve worst case thus raises a second important methodological point: How should robust decision analysis characterize the risks of seemingly unlikely, hard-to-hedge-against futures?

CHARACTERIZING IRREDUCIBLE RISKS

In the standard Bayesian approach, decision analysis begins by eliciting the probabilities of a wide variety of uncertain input parameters from experts and stakeholders. The analysis then provides the strategy with optimum performance contingent on these expectations. The optimum strategy may perform poorly in some futures that the elicitations have suggested are highly unlikely. The strategy's ultimate success may depend strongly on the correct characterization of these risks.

RAP™ takes an opposite approach, consistent with Bayesian theory but inverting Bayesian practice. The iterative process illustrated in Figures 3.1 helps participants design one or several alternative strategies robust across a wide range of plausible scenarios without the need for consensus judgments on the likelihood of these scenarios (although the process can use any reliable, widely accepted probabilistic information available). Each candidate robust strategy may have some small number of futures in which it unavoidably fails. The final step in a robust decision long-term policy analysis is to characterize the risks inherent in adopting a candidate robust strategy. We can make such a characterization by estimating the maximum likelihood that policymakers would have to assign to that strategy's "breaking scenarios" before choosing some other action (including possibly doing nothing). For instance, how likely must scenarios that cause Safety Valve to entail great regret be before policymakers should choose some alternative strategy? Under conditions of deep uncertainty, this approach offers important advantages over the traditional use of probabilities in quantitative decision analysis. Rather than suggest optimal strategies based on experts' estimates of the likelihood of a long list of uncertainties, a robust decision analysis helps decisionmakers craft well-designed strategies that present a small number of irreducible tradeoffs. Policymakers can then focus their attention on these few key uncertainties.

To perform this calculation,[11] the scenario generator produced an ensemble of 2,278 scenarios to span the uncertainty space.[12] Those futures for which the regret of the Safety Valve strategy is high, defined here as greater than 2 percent, were identified. This procedure identified 51 such scenarios, about 2 percent of the total sample. Larger numbers of cases and more sophisticated experimental designs (designed, for example, to oversample in regions of high regret) could increase the number of breaking scenarios, and these measures might be justified in the context of some analyses. This smaller sample is sufficient for the demonstration purposes here.

The next step is to identify the strategy that performed best in those 51 futures where the Safety Valve strategy does poorly. Assuming

---

[11]An early version of this procedure was initially reported in Lempert et al. (2000).

[12]This ensemble resulted from conducting a Latin Hypercube sample of 5,000 points. (Latin Hypercube is a method of sampling that produces values uniformly spread over high-dimensional spaces.) We then kept only those runs whose demographic input parameters produce plausible future population scenarios, defined as a 2050 population between 700 million and 1.5 billion in the North and a 2050 population between 7 billion and 15 billion in the South. We discarded the 2,729 cases that did not meet this criteria.

that all these "breaking scenarios" are equally likely, the strategy that sets and enforces near-term emissions intensity milestones of 1 percent increase in the North and 2 percent increase in the South has the lowest average regret of any option considered. This strategy, labeled M12, does not ask the question "Is this milestone achievable?" shown in Figure 5.12. Unlike Safety Valve, M12 enforces its near-term milestones whether or not they are costly.

The next question is how likely these "breaking scenarios" would need to be in order for policymakers to choose a strategy other than Safety Valve. The 2,278 point uncertainty space sample in this exercise falls into two classes: the $N_{Succeeds}$ = 2,227 points for which Safety Valve has low regret and the N_{Fails} = 51 points for which it does not. Define ϕ as the odds of a breaking scenario relative to a future where Safety Valve succeeds.[13] The expected regret of strategy j can then be written as

$$\overline{E}_j = \frac{N_{Succeeds}\overline{E}_{j,S} + \phi N_{fails}E_{j,F}}{N_{Succeeds} + \phi N_{Fails}} \tag{5.1}$$

where $\overline{E}_{j,S}$ and $E_{j,F}$ are the expected regret of strategy j across the 2,227 and 51 scenarios where Safety Valve succeeds and fails, respectively.[14]

Figure 5.17 compares the resulting expected regrets for Safety Valve and M12, its Best Alternative as a function of the odds that decisionmakers ascribe to a breaking scenario. Both strategies perform poorly when the odds of a breaking scenario are high relative to their performance when such futures are unlikely. However, Safety Valve's expected regret is over twice that of its best alternative when

---

[13]Again, at this stage the analysis seeks to provide decisionmakers with information in a form that will support their final reasoning process. There exist studies indicating that the intuitive appreciation of likelihoods stated in the form of odds is more readily accommodated by most people than in the form of probabilities more familiar to statisticians.

[14]The calculations assume that the scenarios have uniform probability within the $N_{Succeeds}$ and N_{Fails} sets. In general, analysts could calculate the expected regret of strategies within each sample using one or more alternative probability distributions. Similarly, they could also differentiate among different types of breaking scenarios for the strategy of interest and assign different odds to each type.

the odds of a breaking scenario are high. However, the crossover point of the lines in Figure 5.17 suggests that, if decisionmakers are to abandon Safety Valve for its best alternative, they must believe that the odds of any individual breaking scenario are nearly 10 times greater than those of any other future. That is, decisionmakers should choose Safety Valve unless they believe the odds of a future such as that shown in Figures 5.15 and 5.16 are an order of magnitude more likely than that shown in Figures 5.2, 5.3, 5.11, or any of a wide range of other cases.

At present there exists no sound science or any panel of experts who can definitively assign such odds. Rather, decisionmakers, whether

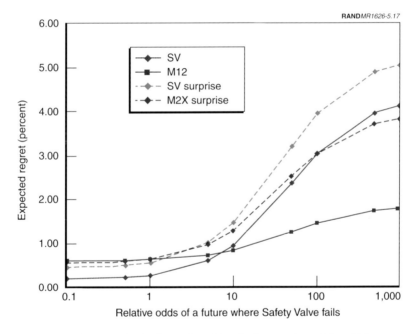

NOTE: The horizontal axis shows the odds of a future where Safety Valve performs worst relative to those where it performs well. The odds are given by ϕ in Equation (5.1). Solid lines show cases with no surprise. Dashed lines show cases with technology surprise described below.

Figure 5.17—Expected Regret of the Safety Valve Strategy and Its Best Alternative in Futures Where Safety Valve Performs Worst

citizens or their representatives, must ultimately judge the likelihood of these key futures through their choice of strategy. Analysis can perform a valuable service in providing a framework for characterizing the irreducible risks of potentially robust strategies. Rather than put pressure on science and experts to seek more certain judgments than are warranted or possible, the robust decision approach applies a wide range of available information to the clever design of a small number of strategies that minimize as much as possible the key uncertainties that must be characterized. In doing so, the approach combines the insights available through data and computer calculation with judgments best left to humans. Decisionmakers are often most comfortable making judgments about how they weigh alternative values and expectations, not in abstract isolation but in the context of the choice among specific alternative policies (Lindblom, 1959). The role of the analyst is to help decisionmakers minimize and characterize as much as possible such key, irreducible tradeoffs. In performing this final characterization of probabilistic risk, we leave to groups of stakeholders a final, informed, and focused judgment of the futures they wish to bet against.

CONFRONTING SURPRISE IN SUSTAINABLE DEVELOPMENT

As described in Chapter Four, this analysis also involved a group of RAND experts from a variety of fields who acted as "stakeholders" in the sustainability debate. We met with the group members individually over several months to familiarize each with the scenario generator, explore the scenario landscapes, and acquaint them with the particular robust decision method and software system being used. The entire group then met to be presented with the argument for Safety Valve as a robust strategy and was then challenged to suggest surprises that might defeat the Safety Valve strategy. This process was intended to match that shown in Figure 3.1.

The group's suggestions are summarized below as three future surprises that could be represented with relatively simple structural changes to the modified Wonderland scenario generator:

- Technology Surprise: Unforeseen technological advances introduce discontinuous change in the nature of the economy and its relation to the environment. This was represented by doubling the exogenous decoupling rate in 2030 and beyond.

- Population Surprise: A global epidemic occurs. This was represented by doubling the death rates between 2030 and 2050.

- Values Surprise: Future generations live in a virtual world of their own creation and care nothing for any of the things familiar to us or our ancestors. We represent this by varying the parameters of future decisionmakers in our model beginning in 2030.

As is often the case, these surprises are best represented by structural rather than parametric uncertainty. That is, referring to the XLRM parsing of the policy problem introduced in Chapter Four, the surprises related not to new values of the parameters of the X quadrant representing factors outside our control (future oil prices, regional wars breaking out, etc.). Rather they occur in the R quadrant, the fundamental relationships tying these factors together. This reflects the fact that the very fabric of the systems we rely upon may also shift.

How might these surprises affect a decisionmaker's views on the robustness of the Safety Valve strategy? Figure 5.18 shows the performance of Safety Valve over the landscape of plausible futures shown in Figure 5.13 for the surprise-free case and each of the three surprises.[15] These surprises do have some small effect on the regret of this strategy as viewed from today. In particular, there are regions where Safety Valve's relative performance degrades compared to other alternatives in the face of the technology surprise. However, the results in Figure 5.18 cast only little doubt on Safety Valve as the robust choice.

These views, of course, represent only a narrow slice across the full range of plausible scenarios. To understand the effects of the surprises across the uncertainty space, we ask how they would change decisionmakers' judgments about the likelihood of futures that would cause them to choose an alternative to the Safety Valve

---

[15]Note that this figure differs from that in Lempert, Popper, and Bankes (2002) because the surprises occur in 2030 rather than 2050.

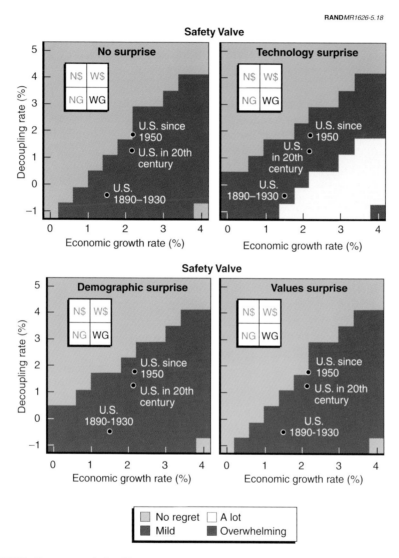

NOTE: Futures are defined by potential surprises identified by our advisory panel: no surprise, a large advance in the decoupling rate, an epidemic that radically reduces population, and a shift in the values of future generations, which causes them to not value environmental quality.

Figure 5.18—Performance of the Safety Valve Strategy over a Range of Surprising Futures

Addressing this question involves repeating the expected regret calculations discussed above for the case where the surprises occur. As
before, the scenario generator produces an ensemble of scenarios
that span the uncertainty space, but this time technology surprises
are allowed to occur.[16] The 5,000-point Latin hypercube experimental design yields 2,264 futures with plausible population trajectories.
Of these, 61 have high regret (greater than 2 percent) for Safety Valve.
Assuming that all these "breaking scenarios" are equally likely, the
strategy M2X, which has no milestone in the South and sets and
enforces a near-term emissions intensity milestone of a 2 percent
increase in the North, emerges as the one with the lowest average
regret of any option considered. Compared to the case without
technology surprises, this Best Alternative to Safety Valve has less
aggressive milestones because the technology surprise can increase
future decoupling rates and thus reduce the need for aggressive
near-term actions.

The dashed lines in Figure 5.17 show the overall effects of the technology surprise on the Safety Valve strategy. The surprise marginally
reduces the performance of the candidate robust strategies so that
one needs only to assume that the breaking scenarios are five times
more likely than Figures 5.2, 5.3, 5.11, or any of a wide range of other
cases.

Nonetheless, this analysis suggests that the surprises proposed by
the advisory group do not significantly challenge the robustness of
the Safety Valve strategy. Undoubtedly, had we conducted a more
effective stakeholder process, given them more time to consider, or
been willing to modify the model to include more extensive effects,
our experts could have succeeded in suggesting surprises that would
have called Safety Valve's robustness into question. Armed with such
proposals, we could have repeated the analysis to determine how
and whether it would be worthwhile to hedge against these proposed
surprises.

---

[16]Specifically, an input parameter indicates that the new scenario generator code
describing a technology surprise should be employed. The Latin hypercube experimental design then includes a sampling over the years (2002 to 2100) in which that
surprise occurs.

Chapter Six

POLICY-RELEVANT LONG-TERM POLICY ANALYSIS

This report demonstrates a methodology for thinking seriously and systematically about near-term steps that can purposefully shape the long-term, unpredictable future. Work in this study and in other efforts (cited in Chapter Three) suggests that the method and technology hold considerable promise for addressing not only problems associated with long-term policy and long-range thinking, but also in application to a wide range of other venues where deep uncertainty inhibits or frustrates more traditional decision support approaches. The promise is there. Nevertheless, it is also appropriate to point out those features that will require further elaboration for this promise to be fully realized in tools widely used by decisionmakers.

The analysis in Chapter Five offers policy recommendations that appear both interesting and plausible, but it falls short of a policy-relevant exercise for two important reasons. First, the scenario generator lacks sufficient resolution to address the actual policy levers of interest to real world decisionmakers and to embrace the full range of scientific, economic, social, and political information available to help adjudicate among alternative strategies. Second, while any robust decision method applied to LTPA is intended as an iterative, interactive process among the stakeholders to a decision, this exercise did not engage such parties. Addressing these challenges raises two additional issues: the need for improved analytic tools for navigating through large, multidimensional scenario spaces and the need for improved treatment of the ways in which current and future generations value the long-term future.

This chapter describes how future work might improve the robust decision methodology implemented here and describes some of the challenges and potential this limited demonstration suggests for the practice of policy-relevant LTPA.

BUILDING POLICY-RELEVANT SCENARIO GENERATORS

Despite its shortcomings, the scenario generator used in Chapter Five was particularly convenient for our demonstration project. Focused on methodology, this proof-of-principle effort worked best based on an existing model. Had the authors created their own scenario generator from scratch, readers might focus undue attention on the idiosyncrasies and complexities of its design rather than on the general demonstration of how *any* scenario generators ought to be used to make arguments about the best policies to shape unpredictable futures. The Wonderland model, the precursor for the scenario generator used here, is also simple enough to allow the easy addition of key elements—e.g., the adaptive response of future generations. Often downplayed in models designed for prediction, such elements are essential to scenario generators designed for LTPA.

Clearly, a policy-relevant exercise, particularly on a topic as expansive and complex as global sustainability, requires greatly improved scenario generators. To be effective, the scenario generators need to explore the full range of policy levers stakeholders and decisionmakers might propose as robust near-term solutions to the sustainability problem, represent the full range of measures that individuals might use to compare the desirability of alternative future scenarios, and exploit the vast array of knowledge relevant to the debates and concerns of those struggling with this policy problem. For example, Wonderland does not represent education, so education levels cannot be included as a measure of a desirable future, and neither can government policies that promote education as a means of achieving robust near-term strategies. Thus, those who believe that education is a crucial element of sustainability could easily challenge the analysis here. In the normal process of employing the iterative method described in Figure 3.1, analysts would address this challenge by augmenting the scenario generator to incorporate the requisite levers and measures. Such augmentation was beyond the scope of this project.

Similarly, the relationships embodied in the Wonderland-based scenario generator, while plausible, do not capture the available range of knowledge and beliefs that could be brought to bear on the sustainability debate. For instance, there exists a vast literature relevant to GSG's claim that their Conventional Worlds scenario, which in essence reflects the so-called Washington Consensus vision of globalization, can easily slide into Barbarization. Policy-relevant LTPA might use multiple scenario generators to capture the full range of relationships among near-term policies, uncertain factors, and measures of desirable futures implicit in today's different schools of economic, demographic, social, and environmental thought and then examine the robustness of alternative near-term policies with these multiple models.

Policy-relevant scenario generators may require much greater resolution than the Wonderland-based system used here. For instance, a sustainability scenario generator should represent many more countries or regions than just an aggregated North and South. Greater regional resolution would allow users to consider the effects on global trends of important differences between such countries as Nigeria, China, and India as well as enable consideration of the desirability of scenarios based on the fates of individual countries. It is crucial to note, however, the reasons for including greater resolution in a scenario generator for LTPA. Traditional computer modelers will often add detail to a model in a futile search for a greater realism that yields more accurate predictions. In contrast, the need to use available information to adjudicate among alternative policy choices drives the resolution needed for LTPA scenario generation. This contrast stems from a difference in directionality. Prediction begins with an attempt to create a true representation of the real world. LTPA begins from the "back end" with the policy question, which in turn drives the design of the scenario generator. Successive iterations of the analysis can produce scenario generators with the ever-higher resolution necessary to answer more-detailed questions about strategies. At each stage, however, the search for robust strategies must drive the choice of information to include.

The ideal scenario generators for LTPA would also be built and operated very differently from most computer models designed for prediction. Most existing models, including Wonderland, assume specific equations that embody the "best" representation of relationships in the real world. In contrast, an analysis seeking robust long-term policy options should aim to capture the full range of dynamic relationships plausible, given current knowledge that, in many cases, is consistent with several alternatively constructed equations. Thus, the scenario generator must not only support exploration across parametric uncertainty but also across alternative systems of equations to explore implications of structural uncertainty as well. The question of whether such a scenario generator is built as one piece of software or as multiple alternatives (an ensemble of scenario generators) is one of implementation and is not crucial here. However constructed, the software supporting LTPA should allow easy exploration over structural uncertainty, ideally during the course of an exercise in response to challenges and hypotheses raised by participants—e.g., "Would you still favor that policy if the economy responded as claimed by Milton Friedman as opposed to Paul Krugman?" Further, the ideal software infrastructure would support real-time extension to the scenario generator.

The pragmatics of software engineering may require that analysts initially design scenario generators to contain a wide range of potentially relevant representations and phenomena, many of which may remain suppressed and unused during any given stage of the analysis. Some new ideas generated by human interlocutors, stimulated by interim results, may always require revision to the software. However, pragmatic engineering constraints imposed by technological limits should not be confused with methodological foundations. LTPA requires that the most diverse ensemble of plausible scenarios be made available for analytic use. This goal is very different from building a single model that is as correct as can be managed.

The International Futures Global Simulation Model

A useful example of the contrast between predictive models and scenario generators can be found in International Futures (IFs), one of today's most frequently used long-range, global simulation models (Hughes, 1999). The project team examined IFs extensively but ultimately chose to not use it for scenario generation. IFs contains impressive levels of geographic, demographic, and economic resolution as it simulates population, food, energy, environmental, economic, and political developments from 1995 through 2100. It simulates trends in 162 countries that may then be aggregated into 14 to 18 fixed regions. These regions are connected by trade, financial flows, and migration as well as by the threats and conflicts arising from international politics. The economic module balances supply, demand, and trade in each of five economic sectors: agriculture, primary energy, raw materials, manufactures, and services.[1]

IFs aims to represent all major processes in the world to forecast the future and investigate how various uncertainties and policy choices can affect that future. Users can create a vast array of alternative scenarios by changing various model input parameters and, in some cases, even by changing the underlying equations. Quite appropriately acknowledged as the current state-of-the-art in its field, IFs has been used widely to investigate the implication of various long-run trends. For instance, the model has been recently and successfully adopted by a major European Union study on the potential impacts of information technology in Europe.

IFs is intended to reflect best available theory rather than any particular world view. Nonetheless, the model embodies key assumptions that limit its utility as the sole base for an LTPA exercise. In particular, IFs' underlying structure assumes that markets and price signals will work effectively to allocate resources throughout the twenty-first century. As such, it embodies the assumptions underlying the GSG Conventional Worlds scenario. But IFs cannot address GSG's concern that these mechanisms might fail.[2] That is, the model's equations cannot represent the market and institutional breakdowns at the heart of the GSG Barbarization scenario. Thus, IFs cannot assess whether or not near-term policies are sufficient to prevent Barbarization because the model cannot produce that future whatever the choice of near-term policies may be.

---

[1]Information on the International Futures (IFs) platform, developed by Barry B. Hughes at the University of Denver is available at http://www.du.edu/~bhughes.

[2]The IFs model was explicitly constructed to address many of the shortcomings of the World3 model used in the Limits to Growth study. The Limits to Growth work was broadly criticized for ignoring the adaptive capacity of human society and, in particular, the potential of the market to take rising prices as a signal to innovate and change behaviors. IFs goes to great effort to model the effects of markets and prices lacking in World3, but the markets in IFs may work too well. The real-world frictional effects of institutions, culture, and human frailty might prove more important than the model assumes.

From Models to Scenario Generators

IFs is but one example of a vast library of existing models often used for scenario analysis but implicitly built from a template originally conceived for prediction. Fortunately, LTPA can employ such models in a variety of ways. First, LTPA might confine itself to policy questions that can be handled by relatively minor modifications of an existing model. For instance, an analysis might use IFs to examine the robustness of alternative near-term policies—for instance, levels of investment in education—across different manifestations of Conventional Worlds. Such an application might require adding code to IFs that would represent the adaptive responses of future decisionmakers, but would otherwise allow the model to be used largely as is. Second, LTPA might create its ensemble of plausible scenarios using multiple existing models, each representing a distinct set of assumptions about the key driving forces in the future world. Compared to other options, this formulation requires less rewriting of computer code but would make it difficult to create sets of scenarios that smoothly interpolate between different views of the future. Finally, analysts might build an entirely new piece of computer code, drawing on the code and data assembled from existing models but explicitly designed for LTPA, whose equations contained a wide variety of alternative and competing representations of the plausible relationships among uncertainties, levers, and outcome measures.

These last two options raise an important question: How should analysts assess the accuracy and utility of a scenario generator for LTPA? A predictive model fails if its forecasts are inconsistent with the future that actually occurs. Based on the practice of the physical sciences, the creator of a predictive model ideally aims to create a system that generates a forecast different from those predicted by competing models and subsequently borne out by future events. To the extent such analyses treat uncertainty, it enters as a perturbation on the main results, blurring the predictions and hence making it more difficult to distinguish conclusively whether or not events falsify one theory or another. This contrasts strongly with the philosophical precepts guiding construction of scenario generators for LTPA. Scenario generators must aim first to embrace the full range of uncertainty to be as expansive as possible, consistent with available information. The ideal scenario generator fails if it cannot be made to represent a plausible future suggested by a party to the policy debate or the LTPA in which it is being used suggests a robust policy that performs poorly in the actual future in a way unanticipated by the analysis.

IMPROVED NAVIGATION

Policy-relevant scenario generators for LTPA will often be more complex, both in run time and number of input parameters, than the simple Wonderland-based system used here. Navigating through larger scenario spaces will require more capable search, sensitivity analysis, and other analytic tools. There exists a vast repertoire of such analytic methods, largely developed for purposes somewhat different than those required for LTPA. In other work, we have begun to review these methods and select those most useful for guiding the search for robust strategies. In particular, we have focused on sensitivity analysis and other statistical methods that can help identify potentially robust strategies and characterize their most important weaknesses.

The current exercise assesses the robustness of alternative strategies by comparing visualizations. For instance, Figures 5.9 and 5.13 suggest that Safety Valve is more robust than No Increase. Such visualizations will always remain important because they can display the relationship among scenarios and clearly illustrate the influence of a small number of uncertainties on a small number of strategies. They are limited, however, in their ability to summarize the performance of many strategies over many dimensions of uncertainty.

Figure 6.1 employs a commonly used type of statistical visualization that can quickly display the comparative performance of strategies.[3] This "box-and-whisker" plot compares the range of regrets over the uncertainty space for four milestone strategies— M00 (No Increase), M11, M22, and MXX—and the Safety Valve.[4] This visualization was produced by calculating the regret of each strategy at each of 5,000 points in the 41-dimensional uncertainty space defined by all the X

---

[3]Our thanks to RAND graduate fellow Joe Hendrickson for performing the calculations in Figures 6.1 and 6.2.

[4]Table B2 defines the No Increase, M11, M22, and MXX strategies.

and L parameters in Table 4.1.[5] (The parameters are described in detail in the appendix.) The particular set of points was created using a Latin hypercube (Iman, Davenport, and Ziegler, 1980) "space-filling" experimental design that samples the entire space as completely and uniformly as possible without undue clustering in any one region. Figure 6.1 shows the range of regrets for each strategy. The upper and lower edges of each box indicate the first and third quartile of the distribution of regrets for each strategy. (The first quartile is indistinguishable from zero regret.) The lines lie 1.5 times the interquartile range past the top of each box, and the dots show individually the 100 highest regrets for each strategy. Such box-and-whisker plots are common, but this is the first application of this display we are aware of for comparing the regrets of strategies.

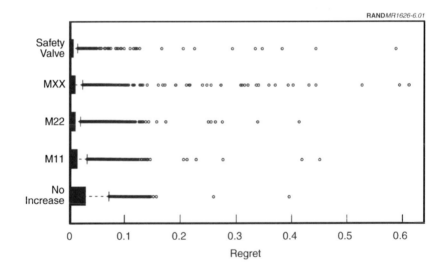

Figure 6.1—Distribution of Regret for Various Milestone and Contingent Strategies

---

[5]Table 4.1 shows 43 uncertainties. Here we held the Future_Action_N and Future_Action_S parameters constant at Decouple+Adapt.

This figure can help users identify potentially robust strategies for further examination. As expected, the Safety Valve strategy has lower regret over most futures compared to the milestone strategies, as measured by the third quartile of their distributions. The best milestone strategy appears to be M22. Interestingly, No Increase has higher regret over many more futures than does the Stay the Course strategy MXX. (MXX has no near-term milestones and thus calls for no near-term policy actions in all futures.) No Increase also has the lowest maximum regret of all the strategies. This result is consistent with the observation that Safety Valve will sometimes allow an otherwise preventable environmental catastrophe. This pattern is also consistent with some of the pathologies of the mini-max criteria originally pointed out by Savage (1950).

While rich in possibilities, Figure 6.1 only provides an initial guide to the robustness of alternative strategies. Implicitly, the figure assumes that all futures are equally important, which might not be the case. To assess robustness, the weaknesses of each strategy must be understood—that is, the futures where each has high regret. In the current exercise, we used a global sensitivity analysis (Saltelli, Chan, and Scott, 2000) to suggest the uncertainties most important in generating high regret for each strategy. This sensitivity analysis helped the project team choose the axes for all the landscape visualizations in Chapter Five, and in particular for Figures 5.10 and 5.14.

Figure 6.2 shows the results of a global, nonlinear sensitivity analysis on the regret of the No Increase and Safety Valve strategies. Global sensitivity analyses are statistical methods that assess the input parameters whose variation is most important in determining variations in outputs, with full consideration of all the interactions among input parameters. The extended Sobol (1990) method was used to conduct these calculations because it is one of the most efficient— i.e., it produces accurate results with the least amount of computation—and because it can employ the same space-filling experimental design as that used for Figure 6.1. (Other sensitivity analysis methods require special space-filling designs less useful for multiple purposes.) The horizontal axis lists 41 input parameters of the Wonderland scenario generator, divided into four categories governing economics, demographics, environment, and actions by future decisionmakers. The vertical axis shows the relative importance of each parameter in producing a large regret for the strategies. Such sensitivity analyses are increasingly common, though the analysis here is novel in one important respect. That analysis is conducted on the regret of strategies, not on the direct model output, which provides information about the parameters most important in determining the choice among strategies, not, as is more common, on the parameters that cause the largest swings in the model outputs.

These sensitivity analysis results suggest that the strategies each have significantly different patterns of weaknesses. The success of the milestone strategies depends most heavily on economic uncertainties, in particular the exogenous rates of decoupling and economic growth and the cost of policies to speed decoupling, and to a lesser extent environmental uncertainties, such as the sustainability threshold. In contrast, the success of Safety Valve strategy depends most strongly on environmental uncertainties and, interestingly, on

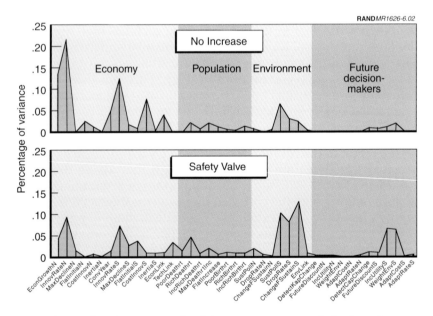

NOTE: Curves show importance of each input parameter to variation in each strategy's regret.

Figure 6.2—Results of Sobol Global Sensitivity Analysis on the No Increase and Safety Valve Strategies

the values and capabilities of future decisionmakers. This information adds richness to the comparison between the two types of strategies because one might regard decoupling rates and economic growth rates as being more readily influenced by near-term policy choices—i.e., by adding additional options to the list of policy levers in Table 4.1—than would be the fragility of the environment or the values of future decisionmakers. Indeed, arguments by those opposed to the Safety Valve proposals for climate change may correspond to the comparative weaknesses of the strategies shown in Figure 6.2. That is, opponents may shun reliance on the actions of future generations in favor of optimism about their ability to sway current technological or economic trends. In a policy-relevant exercise, these results might suggest extensions of the scenario generator and create landscapes that could explore such questions.

The examples shown here only touch on the rich array of analytic and statistical routines that further research might prove useful for policy-relevant LTPA. In future research, we hope to explore the use of these resources, determine those most helpful in such work, and make their use routine.

A DIVERSITY OF MEASURES AND VALUES

To find robust near-term policies that will improve the human condition and that can serve as a basis for a consensus for action, LTPA must employ measures meaningful to a wide variety of individuals and groups. This demonstration employed a set of four measures loosely based on the UN's Human Development Index and concepts of a "Green GDP." The Wonderland-based scenario generator rendered only incomplete version of these indices. More important, the HDI is far from the last word in measuring human progress, as the index's creators would be the first to point out. Policy-relevant LTPA must compare the desirability of alternative scenarios using a set of concise analytic measures that grip the imagination and embrace the goals of many people holding diverse views.

Driven by the conviction that such purely economic indices as GDP present a highly incomplete picture, the quest to develop better indicators of human progress, or lack thereof, has garnered much recent attention. As described in Chapter Four, a growing body of literature has emerged on measures of human progress that correspond to a rich array of human goals. The UN's HDI aims to capture quality-of-life measures, in particular life span and level of educational attainment, in addition to GDP per capita. However, these additional factors tend to be highly correlated with the economic measure. Researchers interested in sustainability have crafted proposed measures for Green GDP, and broader indicators of environmental quality that also address biodiversity and ecosystem services. These attempts have struggled with a lack of key data and uncertainties in the relationships among observable indicators and the overall condition of the environment. Nonetheless, application of Green GDP measures has suggested that some developing nations have achieved apparently high rates of economic growth only by seriously depleting their natural resources. Research on new measures of human progress has been a particular interest at the Pardee Center. For instance, Robert Klitgaard and his colleagues are examining the statistical correlation between self-reported happiness and a variety of databases relating to life expectancy, measures of liberty, and freedom, and they use the results to propose an index of human progress based on the goals of "life, liberty, and the pursuit of happiness."

Additional research on measures of human progress that characterizes the past performance of nations and other entities using actual time series data would be useful. The findings of such research could inform LTPA by suggesting engaging measures that, when adapted as done here for the HDI, can compare the desirability of the alternative, simulated time series created as the scenario generator explores different paths into the future. The robust decisionmaking approach also provides interesting opportunities to address current challenges in creating reliable measures of environmental quality. For instance, proposals to increase the availability of environmental data might become potentially important policy levers to include in strategies. Given the uncertainty in the relationships between observable indicators (e.g., rate of deforestation, extinction rates) and future environmental quality, long-term policy analysts might also explore the types of near-term indicators that, when employed in an appropriate adaptive strategy, can most robustly lead to long-term environmental health.[6]

---

[6]Lempert et al. (2000) provides an initial attempt to examine indicators most useful to a robust strategy.

The great diversity of beliefs and values among those who must participate in confronting many long-term challenges poses a key impediment to generating consensus for action. In our demonstration exercise, a handful of researchers chose alternative weightings for terms in the HDI-derived index to craft four measures of future progress intended to represent the range of worldwide values relevant to sustainability. Policy-relevant LTPA must enable diverse stakeholders to grasp a common view of their challenges and take ownership of the resulting policy recommendations. To that end, LTPA must accomplish two tasks. First, it must partake of the emerging body of research on improved measures of human progress. Second, it must develop and employ group processes with actual communities of stakeholders to create from these innovative new indices an engaging and meaningful set of measures appropriate for each policy challenge.

In addition to employing measures of future human progress that represent the current views of parties to today's decisions, LTPA should also pay careful attention to the evolution of values and worldviews. Many of the most profound historic changes in human society, from the Enlightenment to the spread of democracy and civil rights, resulted in large part from significant changes in how people viewed their world. Near-term actions that influence the evolution of such values may be among the most important levers available to today's policymakers. The GSG argues, for instance, that near-term efforts to change society's current values toward consumption and environment are crucial to enabling the Great Transition scenarios and, thus, long-term sustainability. To address such issues, scenario generators for LTPA must represent the impact of and changes in values over time. The Wonderland-based scenario generator crudely handled the effect of future values by representing future decision-makers as utility-maximizing agents whose preferences and discount rates will be governed by currently uncertain parameters unaffected by any of today's policies. Enriching such descriptions should be a major emphasis of future LTPA research.

ENGAGING THE COMMUNITY OF STAKEHOLDERS

Ultimately, long-term policy analysis aims to influence long-term decisionmaking. Thus the analysis must include and engage the people responsible for such decisions. The robust decision approach to LTPA was conceived in part to address this challenge. The iterative process in Figure 3.1 aims to produce a series of landscapes that allow people to see the relationships among the stories they and others tell about the future and then to seek common solutions that span their differing views. When the results remain counterintuitive or unsatisfactory, people can repeat the process until they can arrive at a common vision about the way to move forward.

In particular, a robust decision method aims to address key shortcomings that previous approaches to LTPA face in engaging communities of stakeholders. The classic image of quantitative policy analysis envisions researchers observing the system of interest with detached objectivity, pursuing intellectual solutions in "ivory tower" isolation from the problem's biases and passions, and presenting final recommendations as completed works. Policymakers and other interested parties must then struggle to fit these recommendations into their multifaceted debates, which generally address a wider range of issues than were contained in the analysis. Practical experience suggests that such detached, deductive approaches do a poor job of eliciting knowledge and opinions from the relevant parties and in promoting a common view among them on the very difficult problems that are the object of LTPA.

Such group processes as Delphi, Foresight, and scenario planning often engage the parties in a debate, but they can fail to subject human deliberations to the stern test of consistency with available information. In particular, such group processes can lack a clear purpose and focus for the exercise, the means for conducting an interpersonal or interorganizational process that will provide a meaningful methodological core without distorting the process itself to fit a predetermined mold, and a way of translating the output and network-building results from the exercise directly into positive plans for action. These lacunae match precisely the aspirations for robust decision methods of LTPA.

This exercise has taken a small but significant step toward fulfilling this vision. The entire LTPA exercise from start to finish was conducted with ongoing interaction, both individually and collectively, with the members of our advisory panel. We addressed some of the challenges they raised and achieved some degree of consensus on the results. In other work, we have also gained experience using these methods to help diverse stakeholders achieve a common vision, albeit in exercises focused on less long-term problems and only traversing a limited number of steps among the pathways laid out in Figure 3.1. For instance, in work supporting strategic planning for the public higher education systems of four U.S. states, landscapes of plausible futures have helped interested parties—including statewide elected officials; chancellors of major university systems; elected governing boards; and business, faculty, and citizen groups— to understand the relationships between different groups' projections of future enrollment and financial pressures and to debate constructively the potential synergies and tradeoffs among the long-term goals favored by different groups (Park and Lempert, 1998). In work for major automobile manufacturers, such visualizations have helped senior managers and design teams to demonstrate that proposed new product plans are robust across the differing forecasts and concerns of diverse parts of their organizations, such as engineering, marketing, and finance.

Two broad steps are required to build from these initial demonstrations the capability for stakeholder interactions necessary for truly policy-relevant LTPA. First, researchers must demonstrate the ability to conduct start-to-finish LTPA exercises with small groups of actual stakeholders who disagree about a policy debate, eliciting their knowledge and concerns, convincing them the landscapes include each of their visions of the future, and moving them toward a common view of necessary policy actions despite their differences in values and beliefs about the future. Second, researchers must show that the common vision produced by such LTPA can engage broad audiences who have not personally participated in the original exercise. Scenario planners, for instance, claim success on both grounds with their well-developed methods for creating scenarios in intensive workshops with stakeholders and then publishing the resulting scenarios to sway a broad audience.

Work to date has laid the foundation for this first step. The XLRM methods described in Chapter Four have proved highly useful in eliciting from decisionmakers and other parties the full range of information—uncertainties, policy levers, multiple measures of the desirability of alternative scenarios, and the relationships among these factors—needed to construct scenario generators for LTPA.

The process in Figure 3.1 can generate visualizations that, with some explanation, engage the participants who spend time in an exercise. Robustness arguments based on such landscapes have demonstrated some ability to dislodge debate participants from entrenched positions and move contending parties toward a more common view.

Only experimentation can determine how best to employ these methods for different policy problems and with different types of groups. For instance, one could conduct parallel workshops each using different types of visualizations presented in different orders. Before-and-after surveys could compare the extent to which the different exercises shaped diverse views into a common vision of action. Another area of research might focus on means to make some small number of scenarios drawn from abstract generated landscapes more concrete. For example, in a sustainability exercise using a scenario generator with the level of regional detail contained in a model resembling IFs, one might imagine drilling down at selected points in a landscape to produce a variety of tables and color-coded world maps, such as those that populate the Human Development Reports, showing the future fate of each country in each scenario. Alternatively or in tandem, one might invent a handful of characters—e.g., three 10-year-olds living in America, India, and Nigeria in 2003—and imagine users drilling down at any point on a landscape to learn the fate of each child and his or her descendants in each scenario. Their fates would be displayed in tables following a few key parameters—e.g., income, health, domicile, number of children, professions, and emigration—images, or perhaps a short computer-generated movie. Only research can determine which, if any, of these techniques make sense under what circumstances.

Less groundwork has been laid to support the second step of disseminating LTPA results so that they influence those who have not participated in a robust decision analytic exercise. Means for doing so might include oversight panels and review boards, distribution of intermediate research products and peer review, workshops and conferences, distribution of software tools to allow distributed collaborative discourse, and community access through web portals.[7]

---

[7]Examples of this approach may be found on the Web at http://www.hf.caltech.edu/ hf/ and in Lempert and Bonomo (1998).

Of course, determining an effective mix of techniques requires much practical experimentation and some systematic research.

IMPROVING LONG-TERM DECISIONMAKING

This report adds robust decisionmaking to the roster of available approaches for LTPA. By reframing the question from "what is the long-term future?" to "how can we shape it to our liking?" this new approach can harness the capabilities of modern computers to grapple directly with the multiplicity of plausible futures that has bedeviled previous approaches. Even using a very simple scenario generator, the demonstration described here suggests a process potentially more complicated and difficult to understand than existing methods. How might robust decisionmaking overcome the challenges of becoming a widely used method of LTPA?

First, perceptions of complication often flow from novelty and unfamiliarity. At first, innovations in any field generally appear more clumsy to use than more established practices. But the balance can change over time. First-year business school students now count among their skills analytic methods once practiced only by a few highly trained specialists. Such familiarity makes tools easier to use but not inherently less complicated. Complication also arises from inadequate infrastructure. A rich body of tools and artifacts now exists to help implement traditional methods for quantitative analysis. The absence of such tools makes robust decision methods for problems of deep uncertainty seem more complicated by comparison. As computers become even more capable, they will provide more fertile ground in which to build and apply such tools. The present research provides a step toward addressing this opportunity.

Second, new decisionmaking tools can influence people well beyond those who actually use them. They can help change the way people think. Decisionmakers often justify policies using conceptual frameworks—optimum choices, cost-benefit analysis—that they could not implement as analytic methodologies. Indeed, the literature holds many examples of highly ingenious and technically sophisticated applications of traditional analytic approaches whose particulars rarely survive first contact with the process of policy formulation. Only a synthesis of the results and a simplified caricature of the original, more-complicated analytic design ultimately influence the decision process.

As one example, the preceding discussion raised the relatively narrow technical issue of how one validates a scenario generator that is not intended to predict the future. A related and deeper question asks what constitutes a proper standard of proof or convincing argumentation when conducting LTPA. The methods discussed in this report depend on a line of argumentation different from the one commonly employed in analytic policy exercises. Traditional approaches depend in large part on principles derived from the scientific method. An extensive literature addresses the nature of scientific arguments and refutations and provides a philosophical base for understanding the nature of scientific proof.[8] Prediction plays a central role in the scientific method, enabling the controlled experiments that will support or refute scientific hypotheses.

This archetype provides a decisive and often unacknowledged model for those who apply analytic techniques to policy problems and the broader audience who judge the relevance of these contributions. This archetype can persist even when considerable evidence suggests its inapplicability to some particular case. For instance, it is widely acknowledged that predictions of the regional impacts of climate change—that is, on particular nations or localities—are highly unreliable. Yet one finds ubiquitous exhortations to produce such predictions as a prelude to decisionmaking. Consider the recent comments of Rajendra Pachauri, chair of the United Nation's scientific body, the Intergovernmental Panel on Climate Change: "I am aware that there is an opportunity for much political debate when you start to predict the impact of climate change on specific regions. But if you want action you must provide this information"(*Nature,* 2003).

This report advocates a new model profoundly simple in concept. Under conditions of deep uncertainty policymakers should identify

---

[8]Some standard works in this literature are Thomas S. Kuhn (1962), *The Structure of Scientific Revolutions*; Karl R. Popper (1962), *Conjectures and Refutations*; and Imre Lakátos, *Proofs and Refutations* (1976). It would also be fruitful, in this connection to consider the commonalities and differences between experimentally based sciences (e.g., physics) and others based more on argumentation from acquired evidence (e.g., paleontology).

near-term strategies robust across a wide range of plausible futures and seek information that alternatively challenges and improves the robustness of alternative policy options. Its full elaboration may be complicated—but not necessarily more so—than any other sophisticated analytic procedure. Human decisionmakers, whether at the individual, organizational, or societal level, commonly confront problems of deep uncertainty with informal reasoning processes similar to the systematic, analytic procedures advocated here. Thus, the results of robust decision-based LTPA should ultimately prove resonant and compelling to broad audiences. By its nature, such a robust-decision approach depends more on open-ended, inductive reasoning than on the conclusive, deductive argument appropriate for policy problems where prediction is feasible. Such lines of argumentation are quite different from what is usually expected from the analytic realm. Unfamiliarity with such types of arguments, and inexperience in disseminating their results to multiple organizations and interest groups helps make these new methods seem complicated. Expanding use and the increasing availability of such tools may ease these problems and help provide a growing analytic foundation for a simple and powerful concept of how decisionmakers ought to think about the long-term.

Chapter Seven

CONCLUSION: MOVING PAST FAMILIAR SHORES

What should people do today to shape the next hundred years to their liking?

Based on experience, a sophisticated reader ought to view with great skepticism the prospect of successfully answering such a question. The checkered history of attempts to predict the long-term future—from the famous declarations that man would never fly, to the *Limits to Growth* study, the unanticipated end of the Cold War, to claims about the "New Economy"—should humble anyone who claims to extend their gaze past the well-charted inner seas on into the deep waters where the future is not tightly constrained by the past. Like navigators sailing before the days of the reliable compass, those who should engage in the vast enterprise of peering into the future often find greater comfort staying focused on the next fiscal quarter, the next year, the next election. They hug that familiar shore where they can feel greater confidence that their predictions need not deviate too far from the known and familiar and still remain credible.

Yet, the long-term future continues to fascinate and beckon—with good reason. The very characteristic that makes it so hard to predict, its relative independence from the constraints of the present, also make it fraught with the greatest dangers and endowed with opportunities. It is only over the course of decades that we can imagine extending freedom and opportunity across a world where vast numbers of people still live with oppression and deprivation. Only as the project of decades can we imagine changing values and technology sufficiently to enable great worldwide wealth to coexist with a rich natural environment. Over that same sweep of time, we can also imagine undermining the rich foundations of democracy, markets, and law, so painfully built up over the centuries, and descending into barbarism and chaos. The biggest paradox is that our greatest potential influence for shaping the future may often be precisely over those time scales where our gaze is most dim. By its nature, where the near-term future is predictable and subject to forces we can quantify today we may have little effect. Where the future is ill-defined, unpredictable, hardest to see, and pregnant with possibility, our actions may well have the largest effects in shaping it.

This report offers an exploration into how it may now be possible to systematically and analytically assess the actions society can take today to shape this long-term future. Given the grand challenges posed by the twenty-first century, the world community needs methods to reconcile our different views the better to define what it is we wish to do and how we can best go about it in a variety of realms that currently resist our attempts at analysis and navigation. In this exploration, the authors offer no way to extend the resolution and depth of our gaze. We suggest no means to change the funda-mental, unpredictable nature of this long-term future. Instead, we present an approach based on capability we did not possess yester-day to answer a fundamentally different question. Rather than pre-dict what the day after tomorrow may bring, we will consider how best, given our persistent inability to predict, we may frame our actions today to shape a future of our liking.

Implementation may be complex in several particulars, yet the basic concept is very simple:

- Computers can help humans create and consider a very large number of plausible long-term futures. This capability means that we now have the tools to create and probe an ensemble of futures sufficiently diverse that it could well include some similar to the one that will actually occur in those aspects important to assessing our choice among strategies.

- Humans can then use the computer to assess which near-term actions perform well, compared to the alternatives, over all these futures using a wide range of values. In general, such actions will attempt to exploit latent opportunities, hedge against unpredictable dangers, and adapt over time as the path into future becomes more clear.

- Humans and computers then search for plausible futures that "break" the chosen strategy. Humans and computer are both very good at such challenges but in different ways. If humans or the computer can find a breaking scenario, they repeat the process. If they still cannot break it, the resulting strategy should support a consensus for successful action.

The outlines of this approach are not new. Humans have probably imagined alternative futures and how their actions help shape those futures for as long as they have been human. In recent decades, computers have helped buttress weaknesses in human decisionmaking, particularly the ability to handle masses of data, ensure consistency with known laws of numbers, behavior, and the physical world, and trace long causal chains. Only in the last few years have computers acquired the power to support directly the patterns of thought and reason humans traditionally and successfully use to create strategies in the face of unpredictable, deeply uncertain futures. In today's era of radical, rapid change; immense possibilities; and great dangers, it is time to harness these new capabilities to help shape our long-term future.

DESCRIPTION OF THE WONDERLAND SCENARIO GENERATOR

Chapter Four uses the "XLRM" framework to organize the information collected for this study's robust decision approach analysis of near-term sustainability policies. The Relationships (R) connect the exogenous uncertainties (X) and policy levers (L) to the metrics (M) used to assess the relative desirability of alternative scenarios. This appendix describes the relationships, uncertainties, and metrics within the XLRM framework. Appendix B describes the policy levers used to construct robust strategies.

RELATIONSHIPS (R)

The relationships in the scenario generator are based on the mathematical equations in Herbert and Leeves's (1998) simple systems dynamics model called Wonderland.[1] The simulation projects future development paths as a function of a wide variety of assumptions about the properties of the future economy, demographics, and the environment. This section reviews its basic equations, while the sections below describe in detail its various parameters, the variation in which represents the exogenous uncertainties, policy levers, and measures.

---

[1] Herbert and Leeves "Troubles in Wonderland," is available at http://journal-ci.csse. monash.edu.au//ci/vol06/herbert/herbert.html. The name Wonderland derives from the pleasant scenarios produced by the model in which per capita income can grow without limit. Troubles descend on Wonderland in scenarios where environmental constraints dominate.

Herbert and Leeves's Wonderland has an economic, population, and environmental module. We made several key modifications to improve the original model's ability to serve as the scenario generator for this study. First, two world regions are included to represent the current developed and currently developing countries. The dynamics of the two regions are coupled through their economic growth rates, their decoupling rates, and the state of their carrying capacities. Additional modifications include introducing an "inertia" that represents a delay in the time it takes for policy interventions to take effect and developing measures that value the output of alternative scenarios. The final addition is a representation of the way in which future generations may respond to concerns about sustainability.

Economy

The economy in region $r = N$ or S of the Wonderland scenario generator is represented by output per capita $Y_{r,t}$ which grows at an exogenous rate γ_r which can be slowed by policy and environmental changes in each region. Thus

$$Y_{r,t+1} = Y_{r,t}\left[1 + \gamma_r - \gamma_{r,t}^{adj} - \varphi_1\gamma_{r',t}^{adj}\right]$$

(A1)

where the drag on economic growth can be caused by decreases in the "carrying capacity" of the environment $K_{r,t}$ and the imposition of any policies aimed at preserving the carrying capacity. We write

$$\gamma_{r,t}^{adj} = \left(\gamma_r + \mu_r^{eff}\right)\left(1 - K_{r,t}\right)^{\lambda_r} + C_r\frac{\tau_{r,t}}{1 - \tau_{r,t}} + C_r^{disconnect}$$

(A2)

where the coefficient C_r scales the amount that innovation policies (represented by the pollution tax $\tau_{r,t}$) may slow growth; μ_r^{eff} is a factor describing the rate at which decreases in the carrying capacity slows economic growth; and $C_r^{disconnect}$ is the cost of policies, described in Equation (A16), that can reduce μ_r^{eff}. The carrying capacity, whose dynamics are described by Equation (A9) below, is an abstraction representing the ability of the natural environment to support economic growth. It ranges in value from an undiminished

1 to a totally depleted 0. The growth in region r can be affected by environmental changes and policy interventions in both the same region, r, and the other region, r′. The latter effect is proportional to the factor $\varphi_1 (0 \leq \varphi_1 \leq 1)$, which characterizes the economic linkages between the regions.

Population

The population in each region grows at a rate proportional to the difference between birth and death rates

$$N_{r,t+1} = N_{r,t} \left[1 + \left\langle \frac{B_{r,t} - D_{r,t}}{1,000} \right\rangle \right],$$ (A3)

where the birth and death rates are functions of regional income and carrying capacity given by

$$B_{r,t} = \beta_0 \left[\beta_r - \left\langle \frac{e^{\beta Y_{r,t}}}{1 + e^{\beta Y_{r,t}}} \right\rangle \right]$$ (A4)

and

$$D_{r,t} = \alpha_0 \left[\alpha_1 - \left\langle \frac{e^{\alpha Y_{r,t}}}{1 + e^{\alpha Y_{r,t}}} \right\rangle \right] \left[1 + \alpha_2 (1 - K_{r,t})^\theta \right].$$ (A5)

We assume that the parameters in Equations (A4) and (A5) governing the dynamics of the population are the same for both regions.

Environment

The annual flow of pollutants in each region is given by the product of regional population, per capita output, and $P_{r,t}$ the pollution per unit output[2]

$$F_{r,t} = N_{r,t} Y_{r,t} P_{r,t}, \tag{A6}$$

where the pollution per unit economic output improves due to the region's exogenous technology innovation χ_r (related to the decoupling rate as shown below) and the cumulative effects of innovation policies in both regions

$$P_{r,t+1} = \left(1 - \tau_{r,t}^{\text{eff}} - \varphi_2 \tau_{r',t}^{\text{eff}}\right) \chi_r P_{r,t}. \tag{A7}$$

The first two factors on the right-hand side of Equation (A7) are related to the policy-induced decoupling rate. The factor $\varphi_2 \left(0 \leq \varphi_2 \leq 1\right)$ represents the extent to which improvements in innovation in one region flow to the other and where the cumulative effect of the innovation policy in region r is given by

$$\tau_{r,t}^{\text{eff}} = \eta_r \tau_{r,t} + \left(1 - \eta_r\right) \tau_{r,t-1}^{\text{eff}}. \tag{A8}$$

The inertia η_r can delay the full effect of the policy and can be an important determinant of the ability of future generations to adapt to sustainability challenges, as discussed below.

The annual flow of pollution can reduce the environmental carrying capacity, which also regenerates itself, according to the expression

---

[2]The Herbert and Leeves Wonderland model also includes a pollution control expenditure that can reduce the annual flow of pollution. We have dropped this factor because it is superseded by the response of future generations described in the next section.

$$K_{r,t+1} = \frac{e^{\ln\left(\frac{K_{r,t}}{1-K_{r,t}}\right)+\delta_r^{\text{eff}}K_{r,t}^{\rho_r}-\omega_r F_{r,t}}}{1+e^{\ln\left(\frac{K_{r,t}}{1-K_{r,t}}\right)+\delta_r^{\text{eff}}K_{r,t}^{\rho_r}-\omega_r F_{r,t}}} \tag{A9}$$

Whether carrying capacity grows or declines over time depends on whether the annual flow of pollution exceeds the sustainable level given by $F_{\text{sustain},r} = \delta_r^{\text{eff}}K_{r,t}^{\rho_r}/\omega_r$. We assume that the level of sustainable pollution in one region can be affected by changes in the carrying capacity in the other region, so that

$$\delta_r^{\text{eff}} = \delta_r\left\{1-\phi_3\left[1-\left\langle\frac{K_{r',t}}{K_{r',0}}\right\rangle^{\rho_{r'}}\right]\right\}, \tag{A10}$$

where the factor $\varphi_3\left(0\le\varphi_3\le1\right)$ represents the extent to which a decline in carrying capacity in one region affects the level of sustainable pollution in the other region.

We limit the carrying capacity to a maximum value of 0.999. If carrying capacity becomes too close to 1, the pollution flow can become what appears to be unrealistically brittle, allowing an overly large overshoot of the sustainable level of pollution and too great a strain on the ability of future generations to respond to changes in the carrying capacity.

Response of Future Generations

One of the most significant changes we made to the Wonderland scenario generator was to add a representation of the response of future generations to sustainability concerns. We assume that future generations will follow the policy set by today's decisionmakers until they detect concrete warning of an imminent sustainability problem. This puts the issue into the form of a two-period, adaptive-decision problem that, although crude, should capture the key features we wish to address.

In the Wonderland scenario generator, a region's carrying capacity grows to or remains at a high level $\left(K_{r,t}\approx1\right)$ until the annual flow of pollution in that region exceeds the sustainable level, as described in Equation (A9). Once the sustainable flow is exceeded, the carrying capacity begins a slow-to-start, but rapidly accelerating decline, whose rate depends on the amount by which the annual flow exceeds the sustainable level. We assume that future (and present) decisionmakers are uncertain of the level of this sustainable flow and other key parameters describing the state of their world until they detect a decrease in their region's carrying capacity. Once such a change is detected, we assume decisionmakers can act with perfect information about these parameters. We assume that the decision-maker can observe changes of Δ_{K_r} or larger, so that the decision-makers in a region start their second, perfect information decision period in the first time interval when $K_{r,t-1} - K_{r,t} > \Delta_{K_r}$.

We assume future decisionmakers act to maximize the expected, present value of their utility as they look out into the future. We write the utility of this future decisionmaker at time t as

$$
U_r(t) = \sum_{\tau=0}^{50} (1-d_r)^\tau N_{r,t+\tau} \left\{ \left(1-W_r^{env}\right) \left[\frac{Y_{r,t+\tau}}{Y_{N,2000}} \right]^{1-\varepsilon_r} + W_r^{env} \left[K_{r,t+\tau} - 1 \right] \right\}, \quad (A11)
$$

where d_r is the discount rate used by decisionmakers in region r, ε_r is their marginal utility of additional income, and W_r^{env} is the weight they put on future changes in the environment as opposed to future increases in income. Both income and environmental terms are normalized with respect to present-day conditions in the developed world.

We assume future decisionmakers can take two actions to maximize their utility as expressed by Equation (A11). They can choose an innovation policy (represented by an annual tax, τ_t, on the pollution flow), which, as shown in Equations (A6), (A7), and (A8) and Equation (A2), respectively, reduces the annual flow of pollution at the cost of slowing economic growth. In addition to any choice about the tax, future decisionmakers can also choose to invest in actions that will to some extent reduce the effects of environmental decline on economic growth. If such an investment is made at time t', the environmental decline parameter in Equation (A2) is given by

$$\mu_{r,t}^{eff} = \left(1 - r_r^{disconnect}\right)^{t-t'} \mu_r,$$ (A12)

where $r_r^{disconnect}$ is the rate at which dependence on the environment is reduced and μ_r is the rate of decline in the absence of any policy intervention. The cost of such investments, $c_r^{disconnect}$, can slow economic growth as shown in Equation (A2).

Every five years, starting in the period they first detect a decline in carrying capacity, future decisionmakers can adjust the tax level or choose investments that reduce the effect of any collapse of environmental carrying capacity on subsequent economic growth. However, we assume that such investments are irreversible—that is, once made, they cannot be undone. In calculating the utility resulting from alternative choices, we assume that the decisionmakers in each region assume that the decisionmakers in the other region will continue their then-current policies indefinitely into the future.

UNCERTAINTIES (X)

The exogenous uncertainties are represented by the range of plausible values for the input parameters to Wonderland. The CARs control panel for Wonderland lists the following parameter names to provide values for the factors in Equations (A1–A12):

Economic Parameters

Base_Economic_Growth_Rate_N is the exogenous rate of economic growth given by the γ_N parameter in Equation (A1).

Decoupling_Rate_N is the rate at which technology innovation reduces pollution flow per unit output in the North. It is related to the innovation rate χ_N in Equation (A7) by $\chi_N = 1 - \text{Decoupling\_Rate\_N}$.

Convergence_Year is the year in which per capita income in the South reaches parity with that in the North. It is related to the γ_S parameter in Equation (A1) by

$$\gamma_S = \left(1 + \gamma_N\right) 6.93^{1/(\text{Convergence\_Year}-2000)} - 1,$$

where 6.93 is the current ratio of income in North and South.

Del_Decouple_Rate_S is the difference in the decoupling rates in North and South. It is related to the innovation rate χ_S in Equation (A7) by $\chi_S = \chi_N - \text{Del\_Decoupling\_Rate\_S}$.

Max_Rate_of_Decline is the maximum rate at which a decline in natural capital can reduce economic growth given by the μ parameter in Equation (A2). There are separate values for each region.

Flatness_of_Initial_Decline governs the steepness of the initial drop in economic growth, given by the exponent λ in Equation (A2). There are separate values for each region.

Cost_to_Speed_Decoupling governs the effect of innovation policies (represented by pollution taxes) on growth, given by the C parameter in Equation (A2). There are separate values for each region.

Inertia governs the time it takes for pollution taxes to achieve their full effect on the flow of pollution. The η parameter above is given by $\eta = \ln(50)/\text{Inertia}$. Inertia ranges from 0 to 50 years with a nominal value of 20 years. There are separate values for each region.

Econ_link governs the extent to which economic changes in one region affect the other, given by the φ_1 parameter in Equation (A1).

Tech_link governs the extent to which innovation rate changes in one region affect the other, given by the φ_2 parameter in Equation (A7).

Population Parameters

Poor_Deathrate is the population's death rate at small output per capita.

Rich_Deathrate is the population's death rate at high output per capita.

Income_for_Rich_Deathrate is the output needed to yield the Rich_Deathrate.

The Wonderland parameters in Equation (A5) are related to these parameters by the following:

$$\alpha = \ln\left[\frac{1-.01}{.01}\right] \bigg/ \text{Income\_for\_Rich\_Deathrate}$$

$$\alpha_0 = 2\big(\text{Poor\_Deathrate} - \text{Rich\_Deathrate}\big)$$

$$\alpha_1 = \big(2\text{Poor\_Deathrate} - \text{Rich\_Deathrate}\big) \bigg/ \alpha_0.$$

Max_Deathrate_Increase is the maximum increase in death rate due to a decline in natural capital, given by the α_2 parameter in Equation (A5).

Flatness_of_Increase governs the steepness of the initial increase in death rate as natural capital first begins to decline, given by the θ exponent in Equation (A5).

Poor_Birthrate is the birthrate at zero output per capita.

Rich_Birthrate is the birthrate at high output per capita.

Income_for_Rich_Birthrate is the output needed to yield the Rich_Birthrate.

The Wonderland parameters in Equation (A4) are related to these parameters by the following:

$$\beta = \ln\left[\frac{1-.01}{.01}\right] \bigg/ \text{Income\_for\_Rich\_Birthrate}$$

$$\beta_0 = 2\big(\text{Poor\_Birthrate} - \text{Rich\_Birthrate}\big)$$

$$\beta_1 = \big(2\text{Poor\_Birthrate} - \text{Rich\_Birthrate}\big) \bigg/ \alpha_0.$$

Environment Parameters

Sustainable_Pollution is the sustainable level of pollution, $F_{sustain}$. There are separate values for each region.

Rate_of_Drop is the rate at which natural capital falls when pollution flows are above the sustainable level. Specifically, it is the rate of decline when natural capital is equal to 0.99 and the pollution flow is 10 percent above $F_{sustain}$. There are separate values for each region.

Change_in_F_Sustain is the change in $F_{sustain}$ when natural capital is cut in half. When it is less than one, $F_{sustain}$ drops with declining natural capital. When it is greater than one, $F_{sustain}$ increases with declining natural capital. There are separate values for each region.

The Wonderland parameters in Equation (A9) are related to these parameters by the following:

$$\rho = \frac{\ln\left(Change\_in\_F\_Sustain\right)}{\ln\left(0.5\right)}$$

$$\delta = \frac{1}{2.1}\ln\left[99\,Rate\_of\_Drop\right]$$

$$\omega = \frac{\delta}{Sustainable\_Pollution}.$$

Env_link governs the extent to which environmental changes in one region affect the other, given by the φ_3 parameter in Equation (A10).

Future Response

Future_Action specifies the type of future response in region r. It can specify no future response (No Response), a future response based on innovation policies that increase decoupling rates only (Decoupling), and a future response that can employ both decoupling as well as investments that reduce any effects of environmental degradation on economic growth (Decouple+Adapt). There are separate values for each region.

Detect_Capital_Change, Δ_K, is the minimum decline in natural capital detectable by the adaptive decision strategy. There are separate values for each region.

Future_discount, d, is the discount rate used by future decisionmakers in calculating the utility they will get from alternative policies. There are separate values for each region.

Income_utility, ε, is the marginal utility future decisionmakers get from an increase in output per capital. There are separate values for each region.

Weight_env, W^{env}, is the weight future decisionmakers place on environmental change as opposed to income increases. There are separate values for each region.

Table A.1 lists the range of values we used for each of these parameters in this study. For each parameter the table lists the nominal value used for all the visualizations in this study, except where otherwise specified. The table also lists the low and high values used to constrain the searches and the experimental designs used for the sensitivity analyses. For some parameters, these high, low, and nominal values were chosen based on available data. For instance, the discussion surrounding Figure 5.1 provides our estimates of plausible ranges for the economic growth and decoupling rate parameters. Where data was not readily available, it was sufficient for the purposes of this demonstration exercise to take nominal values from Herbert and Leeves (1998) and arbitrarily choose high and low values to define the parameter ranges.

MEASURING OUTCOMES (M)

In this report we also added to the original Wonderland model measures to compare the desirability of alternative long-term futures.[3] The first is a Quasi-HDI measure, a weighted, discounted average of the annual improvement in four factors: net output per capita (that

---

[3]These measures are actually described in the Wonderland "run method," the computer code that connects the simulation to the CARs software environment.

Table A.1

Uncertain Parameters in Wonderland Scenario Generator

Parameter	North	South
Economy		
Base_Economic_Growth_Rate_N	0%, 1.5%, 4%	NA
Decoupling_Rate_N	–1%, 1.5%, 5%	NA
Convergence_Year	NA	2050, 2140, 2400
Del_Decouple_Rate_S	NA	–3%, 1%, 3%
Max_Rate_of_Decline	0%, 3%, 16%	0%, 3%, 16%
Flatness_of_Initial_Decline	0, 2, 4	0, 2, 4
Cost_to_Speed_Decoupling	0, 0.001, 0.01	0, 0.001, 0.01
Inertia	5, 30, 50 years	5, 30, 50 years
Econ_link	0, 0.5, 1	NA
Tech_link	0, 0.5, 1	NA
Population		
Poor_Deathrate	15, 20, 25	NA
Rich_Deathrate	5, 15, 30	NA
Income_for_Rich_Deathrate	$5, $10, $50	NA
Max_Deathrate_Increase	0, 1, 10	NA
Flatness_of_Increase	0, 15, 100	NA
Poor_Birthrate	30, 35, 40	NA
Rich_Birthrate	5, 15, 20	NA
Income_for_Rich_Birthrate	$10, $40, $100	NA
Environment		
Sustainable_Pollution	0, 2, 10	0, 2, 10
Rate_of_Drop	0, 0.005, .01	0, 0.0005, 01
Change_in_F_Sustain	0.5, 0.871, 1.3	0.5, 0.871, 1.3
Env_link	0, 0, 1	NA
Future Decisionmakers		
Future_Action	No Response, Decouple, Decouple+Adapt	No Response, Decouple, Decouple+Adapt
Detect_Capital_Change	0%, 1%, 10%	0%, 1%, 10%
Future_discount	0%, 2%, 10%	0%, 2%, 10%
Income_utility	0, 0, 1	0, 0, 1
Weight_environment	0, 0, 1	0, 0, 1
Disconnect_cost	0, 0.2, 10	0, 0.2, 10
Disconnect_rate	0, 0.1, 0.2	0, 0.1, 0.2

NOTE: Triplets of numbers show low, nominal, and high parameter values. (For Future_Action, the nominal value is Decouple+Adapt.)

is, output minus environmental control costs), death rates, the annual flow of pollution, and the carrying capacity. The second is a "Worst Decade" measure, a weighted average of the worst 10-year decrease in the net output per capita, death rate, and pollution flow time series. Each measure is a weighted average of each time series in each region.

The Quasi-HDI measure is given by

$$
\text{Quasi-HDI}_\tau =
$$
$$
W_N \sum_{t=0}^{\text{Last\_Year}} (1-d)^t \text{Index}_{N,t+\tau} + (1-W_N) \sum_{t=0}^{\text{Last\_Year}} (1-d)^t \text{Index}_{S,t+\tau}, \quad \text{(A13)}
$$

where the index for each region is given by

$$
\text{Index}_r =
$$
$$
W_1 \left(\frac{Y_{r,t}}{Y_{r,t-1}} - 1 \right) + W_2 \left(\frac{D_{r,t-1}}{D_{r,t}} - 1 \right) + W_3 \left(\frac{F_{r,t-1}}{F_{r,t}} - 1 \right) + W_4 \left(\frac{K_{r,t}}{K_{r,t}} - 1 \right). \quad \text{(A14)}
$$

The first term is proportional to changes in income per capita. The second, death rate term is a proxy for lifespan in the actual HDI measure. The two pollution terms allow an environmental component, which can depend on the annual flow of pollution or the state of the carrying capacity. The Worst Decade measure uses the same components as the Quasi-HDI measure and is given by

$$
\text{Crash}_t = W_N \text{Index}_{N,t} + (1-W_N) \text{Index}_{S,t}, \quad \text{(A15)}
$$

where the index for each region is given by

$$
\text{Index}_{r,t} =
$$
$$
W_1 \min_{t<T<\text{Last\_Year}} \left[\left(\frac{Y_{r,T}}{Y_{r,T-10}} \right)^{1/10} - 1 \right] + W_2 \min_{t<T<\text{Last\_Year}} \left[\left(\frac{D_{r,T}}{D_{r,T-10}} \right)^{1/10} - 1 \right] \quad .\text{(A16)}
$$
$$
+ W_3 \min_{t<T<\text{Last\_Year}} \left[\left(\frac{F_{r,T-10}}{F_{r,T}} \right)^{1/10} - 1 \right] + W_4 \min_{t<T<\text{Last\_Year}} \left[\left(\frac{K_{r,T}}{K_{r,T-10}} \right)^{1/10} - 1 \right]
$$

The CARs control panel for the Wonderland Scenario generator lists the following parameter names to provide input to Equations (A13–A16):

Measure specifies the choice of output measure, either Quasi-HDI or Crash.

Vantage is the year from which the measures are assessed. It can take values of "Today," "Milestone Year," "Year Adaptive Begins," 2025, and 2050.

Weight_North is the weight W_N placed on the North as opposed to South outputs.

Weight_Output is the weight W_1 on the output per capita term.

Weight_LifeSpan is the weight W_2 on the death rate term.

Weight_Env is the weight W_3 on the pollution term.

Weight_Capacity is the weight W_4 on the carrying capacity term.

Discount_Rate is the discount rate d used in the HDI-like measure.

Last_Year is the last year considered by the HDI-like and crash measures.

In this study, we used the Quasi-HDI measure to describe four different value systems—North Quasi-HDI, World Quasi-HDI, North Green-HDI, World Green-HDI—that roughly characterize the values of the diverse participants in the sustainability debate. Table A.2 shows the parameters used to describe each of the value systems.

Table A.2

Parameter Values Defining Four Measures Used in This Report

Measure	North Quasi-HDI Quasi-HDI	World Quasi-HDI Quasi-HDI	North Green-HDI Quasi-HDI	World Green-HDI Quasi-HDI
Last_Year	100	100	100	100
Vantage	Today	Today	Today	Today
Weight_North	1.0	0.2	1.0	0.2
Weight_Output	1.0	1.0	1.0	1.0
Weight_LifeSpan	1.0	1.0	1.0	1.0
Weight_Env	0.0	0.0	0.0	0.0
Weight_Capacity	0.0	0.0	1.0	1.0
Discount_Rate	2%	2%	2%	2%

ASSESSING ROBUST STRATEGIES

This appendix provides more details on the assessment of robust strategies presented in Chapter Five.

REPRODUCING THE GSG SCENARIOS

The "Overview" section of Chapter Five reproduces the three Global Scenario Group (GSG) scenarios using the Wonderland scenario generator. Table B.1 lists the input parameter values that generate the time trajectories shown in Figure 5.2. The table divides the parameters into four categories: economy, population, environment, and the response of future decisionmakers. We chose the values used here through manual experimentation, starting with the nominal parameters values presented in Appendix A and making the minimal changes needed to reproduce as best as possible the economic and population trajectories shown in Figure 2.1. There is no claim that these are a unique set of parameters that reproduce the GSG scenarios.

The italicized entries in the table show parameters that differ among the three scenarios. In particular, Conventional World and Barbarization have faster overall economic growth rates, but the South reaches economic parity with the North 40 years earlier in Great Transitions. The exogenous decoupling rates in Barbarization are slower than in the other two futures. The rates at which an environmental collapse drags down the economy, Max_Rate_of_Decline, is higher in Barbarization than in the other two cases.

Table B.1

Parameters Describing GSG Scenarios with Wonderland Scenario Generator

	Conventional World	Barbarization	Great Transition
Economy			
Base_Economic_Growth_Rate_N	*1.80%*	*1.80%*	*1.20%*
Decoupling_Rate_N	*1.50%*	*1.10%*	*1.50%*
Max_Rate_of_Decline_N	*3%*	*4%*	*3%*
Flatness_of_Initial_Decline_N	2	2	2
Cost_to_Speed_Decoupling_N	0.001	0.001	0.001
Inertia_N	30	30	30
Convergence_Year	*2140*	*2140*	*2100*
Del_Decouple_Rate_S	1%	1%	1%
Max_Rate_of_Decline_S	*3%*	*4%*	*3%*
Flatness_of_Initial_Decline_S	2	2	2
Cost_to_Speed_Decoupling_S	0.001	0.001	0.001
Inertia_S	30	30	30
econ_link	50%	50%	50%
tech_link	50%	50%	50%
Population			
Poor_Deathrate	20	20	20
Rich_Deathrate	*15*	*15*	*10*
Income_for_Rich_Deathrate	10	10	10
Max_Deathrate_Increase	1	1	1
Flatness_of_Increase	15	15	15
Poor_Birthrate	35	35	35
Rich_Birthrate	*15*	*15*	*10*
Income_for_Rich_Birthrate	*40*	*40*	*17*
Environment			
Sustainable_Pollution_N	*2.00*	*1.25*	*2.00*
Drop_Rate_N	0.0005	0.0005	0.0005
Change_in_F_Sustain_N	0.871	0.871	0.871
Sustainable_Pollution_S	*2.00*	*1.50*	*2.00*
Drop_Rate_S	0.0005	0.0005	0.0005
Change_in_F_Sustain_S	0.871	0.871	0.871
env_link	*0%*	*50%*	*0%*
Future_Response			

Table B.1—continued

	Conventional World	Barbarization	Great Transition
Future_Action_N	Decouple+ Adapt	Decouple+ Adapt	Decouple+ Adapt
Detect_Capital_Change_N	1.0%	10.0%	1.0%
Future_Discount_N	2.0%	10.0%	2.0%
Income_Utility_N	0	0	0.9
Weight_Env_N	0	0	0.5
Adaptation_Cost_N	0.20	1.00	0.20
Adaptation_Rate_N	0.10	0.02	0.10
Future_Action_S	Decouple+ Adapt	Decouple+ Adapt	Decouple+ Adapt
Detect_Capital_Change_S	1.0%	10.0%	0.1%
Future_Discount_S	2.0%	10.0%	2.0%
Income_Utility_S	0	0	0.9
Weight_Env_S	0	0	0.5
Adaptation_Cost_S	0.20	1.00	0.20
Adaptation_Rate_S	0.10	0.02	0.10

NOTE: Italicized lines indicate parameters that differ among scenarios.

The population parameters in Great Transitions differ from the other two scenarios, showing lower birth and death rates in richer societies, and a lower level of wealth needed before the transition to these lower rates.

The environment is more fragile in Barbarization, with lower levels of sustainable pollution in the North and South and stronger coupling between the carrying capacity in the two regions.

The values and capabilities of future decisionmakers also differ strongly among the three cases. In Conventional World and Great Transitions, decisionmakers in both North and South have good ability to detect environmental damage and have sufficiently low discount rates that they react to information about oncoming problems. In contrast, future decisionmakers in Barbarization have little detection capability and heavily discount the future. They thus have little ability or inclination to respond to any sustainability challenges that confront them. Future decisionmakers in Great Transitions hold different values, described by utility functions which more heavily weight the environment as per-capita income grows compared to decisionmakers in the other two cases. Finally, decisionmakers in Conventional World and Great Transitions have an option of adapting to deleterious environmental changes that is not available to decisionmakers in Barbarization.

FIXED NEAR-TERM STRATEGIES

The Wonderland scenario generator provides today's decisionmakers one basic action they can take to influence the development of the economy, environment, population, and ultimately the situation in which future decisionmakers find themselves. Today's decisionmakers can set an annual tax, τ_t, on the pollution flow. The robust decision approach in Chapter Five describes a variety of ways in which today's decisionmakers can prescribe the evolution of these tax rates, one in the North and one in the South, over time in their attempt to ensure long-term sustainability.

The section "Landscapes of Plausible Futures" in Chapter Five assesses the most basic type strategy, one that specifies today the precise path each tax rate will take into the future. We call these strategies "fixed" because they seek to encourage some specific improvement in the exogenous decoupling rate. We consider 16 such fixed strategies across the three GSG scenarios and a wide range of other plausible futures. These strategies differ in their near-term prescribed pollution tax policy. The annual pollution tax $\tau_{r,t}$ as shown in Equation (A8) in each region from the present (2002) through the time $T_{\text{init,r}}$ is given by

$$\tau_{r,t} = \tau_{\text{init,r}} \frac{t - 2002}{T_{\text{init,r}} - 2002},$$

(B1)

where decisionmakers choose today the Initial_Tax, $\tau_{\text{init,r}}$, the maximum pollution tax and the Tax_Date, $T_{\text{init,r}}$, the date that the pollution tax reaches its maximum level. After $T_{\text{init,r}}$ the tax in each region stays constant at the value $\tau_{\text{init,r}}$, until and if future decisionmakers decide to adopt a different policy. The values of the initial tax and period when it goes fully into effect are listed in Table B.2. In particular, Figures 5.2 and 5.3 show the effects, respectively, of the Stay the Course and Slight Increase strategies on the three GSG scenarios.

Table B.2

Fixed Near-Term Strategies

Innovation Policies	$\tau_{init,N}$	$\tau_{init,S}$	$T_{init,N}$	$T_{init,S}$
Stay the Course	0%	0%	2010	2010
NS01	0%	1%	2010	2010
NS02	0%	2%	2010	2010
NS04	0%	4%	2010	2010
NS10	1%	0%	2010	2010
Slight Increase	1%	1%	2010	2010
NS12	1%	2%	2010	2010
NS14	1%	4%	2010	2010
NS20	2%	0%	2010	2010
NS21	2%	1%	2010	2010
NS22	2%	2%	2010	2010
Crash Effort	2%	4%	2010	2010
NS40	4%	0%	2010	2010
NS41	4%	1%	2010	2010
NS42	4%	2%	2010	2010
NS44	4%	4%	2010	2010

FINDING WORST CASES FOR THE NO INCREASE MILESTONE STRATEGY

After finding that none of the fixed strategies in Table B.1 are robust, Chapter Five next explored a class of strategies that set a near-term goal and then adjusted policies adaptively over time to reach the goal. The annual pollution tax in each region from the present through the time $T_{init,r}$ is given by

$$
\tau_{r,t} = \begin{cases} \dfrac{\Delta_r}{\left[1-\left(1-\eta_r\right)^{T_{init,r}-2000}\right]\left(1-\chi_r\right)} \dfrac{t-2002}{T_{init,r}-2002} & \text{if } \Delta_r \geq 0 \\ 0 & \text{if } \Delta_r < 0 \end{cases}, \text{(B2)}
$$

where $\Delta_r = \gamma_N - \text{Decoupling\_Rate\_N\_Goal}$ and $\Delta_S = \gamma_S - \text{Decoupling\_Rate\_N} - \text{Del\_Decouple\_Rate\_S\_Goals}$. When used in Equations (A7) and (A8), $\tau_{r,t}$ produces a decoupling rate at the desired goal by the desired time. Today's decisionmakers choose the goals and the date the goals will be met. The annual tax then adjusts itself based on the revealed values of economic growth, exogenous decoupling rates, and inertia. After $T_{init,r}$ the tax stays constant at its $T_{init,r}$ value until decisionmakers decide to adopt a different policy.

Table B.3 shows the parameters that define these 16 milestone strategies. Note that in contrast to the fixed strategies, a smaller number here indicates a more aggressive strategy because it indicates a smaller allowed deviation between the rate of decoupling and economic growth—that is, the expression in the parenthesis in the numerator of Equation (B1).

"Exploring Near-Term Milestones" in Chapter Five explored the robustness of the No Increase milestone strategy. In particular, we conducted a computer search to find the future with the largest regret for this strategy. A vast literature on global optimization and many algorithms are available to help find such extreme cases. In this study, it was sufficient to conduct a relatively crude but simple procedure. First, a "downhill" search was launched at each of 500 randomly chosen starting points in the ensemble of plausible futures. The intermediate steps in each downhill search were kept, rather than discarded, which yielded 1,502 futures. Next, those runs

Table B.3

Milestone Strategies Considered in This Report

Goal Strategies	Goal_N	Goal_S	$T_{init,N}$	$T_{init,S}$
No Increase	0%	0%	2010	2010
M01	0%	1%	2010	2010
M02	0%	2%	2010	2010
M04	0%	none	2010	2010
M10	1%	0%	2010	2010
M11	1%	1%	2010	2010
M12	1%	2%	2010	2010
M14	1%	none	2010	2010
M20	2%	0%	2010	2010
M21	2%	1%	2010	2010
M22	2%	2%	2010	2010
M24	2%	none	2010	2010
M40	none	0%	2010	2010
M41	none	1%	2010	2010
M42	none	2%	2010	2010
Stay the Course	none	none	2010	2010

whose input parameters produced what we regarded as implausible demographic futures were eliminated. We defined implausible population trajectories as a 2050 population less than 700 million or greater than 1.5 billion in the North and less than 7 billion or greater than 15 billion in the South. A total of 665 of the original cases met this criteria. Finally, the case with the largest No Increase regret, using any of the four value systems defined in Table A.2, was chosen from these remaining 665 futures. Table B.4 shows the resulting worst-case future for the No Increase strategy. This worst case occurs for the World Green-HDI value system.

Table B.4

Parameters Describing No Increase and Safety Valve Worst Cases

	No Increase Worst Case	Safety Valve Worst Case
Economy		
Base_Economic_Growth_Rate_N	4.0%	3.9%
Decouple_Rate_N	–0.7%	1.9%
Max_Rate_of_Decline_N	0.063	0.068
Flatness_of_Initial_Decline_N	0.266	1.134
Cost_to_Speed_Decoupling_N	0.00012	0.005
Inertia_N	42.9	14.8
Convergence_Year	2059	2087
Del_Decouple_Rate_S	2.4%	–2.9%
Max_Rate_of_Decline_S	0.011	0.142
Flatness_of_Initial_Decline_S	3.876	3.938
Cost_to_Speed_Decoupling_S	0.007	0.008
Inertia_S	49.5	8.3
econ_link	99.0%	16.1%
tech_link	92.8%	67.2%
Population		
Poor_Deathrate	17.14	17.435
Rich_Deathrate	20.363	23.29
Income_for_Rich_Deathrate	23.117	42.607
Max_Deathrate_Increase	1.859	9.369
Flatness_of_Increase	81.294	13.933
Poor_Birthrate	38.942	35.074
Rich_Birthrate	17.089	8.3
Income_for_Rich_Birthrate	24.939	90.94

Table B.4—continued

	No Increase Worst Case	Safety Valve Worst Case
Environment		
Sustainable_Pollution_N	7.54	9.52
Drop_Rate_N	*0.001*	*0.004*
Change_in_F_Sustain_N	1.02	0.57
Sustainable_Pollution_S	8.49	6.13
Drop_Rate_S	*0.004*	*0.006*
Change_in_F_Sustain_S	0.95	1.15
env_link	12%	97%
Future_Response		
Future_Action_N	Decouple+Adapt	Decouple+Adapt
Detect_Capital_Change_N	9.4%	3.2%
Future_Discount_N	*1.6%*	*9.1%*
Income_Utility_N	0.71	0.36
Weight_Env_N	0.51	0.30
Adaptation_Cost_N	4.73	4.91
Adaptation_Rate_N	0.19	0.07
Future_Action_S	Decouple+Adapt	Decouple+Adapt
Detect_Capital_Change_S	5.1%	*4.3%*
Future_Discount_S	4.6%	5.5%
Income_Utility_S	0.34	0.56
Weight_Env_S	0.08	0.90
Adaptation_Cost_S	9.59	4.80
Adaptation_Rate_S	0.12	0.01

NOTE: Italics indicate parameters identified as most important by sensitivity analysis.

The italicized entries in Table B.4 correspond to the parameters found to be most important in the sensitivity analysis described in "Improving Long-Term Decisionmaking" in Chapter Six. In particular, economic growth rates, the exogenous decoupling rate, the strength of linkages between the Northern and Southern economies, and the cost of policies to speed decoupling in the South are important to the regret of the No Increase strategy and are exceptionally high compared to their nominal values in this strategies' worst case. The Southern environment in this worst case is also highly sensitive compared to the nominal cases. Combined with Figure 5.10, this information suggests that the No Increase strategy fares poorly in those futures where exogenous rates of economic growth and decoupling, combined with the high costs of speeding decoupling, make it difficult to reach the milestone.

FINDING WORST CASES FOR THE SAFETY VALVE STRATEGY

After finding significant breaking scenarios for the No Increase strategy in Table B.3, Chapter Five next explores the performance of a Safety Valve strategy, which sets a near-term goal and a cost threshold and then adjusts policies to reach that goal to the extent possible without exceeding the cost threshold. Under Safety Valve, the annual pollution tax in each region from the present through the time $T_{init,r}$ is given by

$$\tau_{r,t} = \begin{cases} \tau_{r,t}^{eff} \text{ if } C_r\left[\tau_{r,t}^{eff}\right]^2 \leq \text{Limit}_r \\ \\ \dfrac{\text{Limit}}{\varepsilon_r} \text{ if } C_r\left[\tau_{r,t}^{eff}\right]^2 > \text{Limit}_r \end{cases}, \tag{B3}$$

where

$$\tau_{r,t}^{eff} = \begin{cases} \dfrac{\Delta_r}{\left[1-\left(1-\eta_r\right)^{T_{init,r}-2002}\right] * \left(1-\chi_r\right)} \dfrac{t-2002}{T_{init,r}-2002} & \text{if } \Delta_r \geq 0 \\ \\ 0 & \text{if } \Delta_r < 0 \end{cases} \tag{B4}$$

Today's decisionmakers choose the near-term goals, Goal_r, the date they will be met, $T_{init,r}$, and the cost limit, Limit_r. The annual tax then adjusts itself to meet the prescribed goal at the appointed time given the revealed values of economic growth, exogenous decoupling, and inertia, unless the costs required in any year exceed the cost threshold, in which case the tax is constrained by the cost target. As with the other strategies, the tax stays constant after $T_{init,r}$ until future decisionmakers decide to adopt a different policy.

Table B.5 shows the parameters values that define this Safety Valve strategy and the milestone strategies to which it is compared.

"Identifying a Robust Strategy" in Chapter Five examines the robustness of the Safety Valve strategy. We find that this strategy is considerably more robust than No Increase. For instance, Figure B.1 shows the performance of Safety Valve over the same set of futures as shown for No Increase in Figure 5.10. The former strategy performs significantly better.

Nonetheless, Safety Valve can still fail in some futures. Similar to the procedure followed for No Increase, we conducted a Latin hypercube sample to find the strategy's worst case. We initially considered a

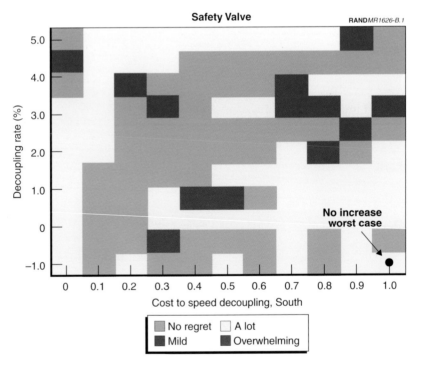

NOTE: These futures are the same as those shown in Figure 5.14. The regret is calculated in comparison to 16 alternative milestone strategies.

Figure B.1—Performance of Safety Valve Strategy over a Landscape of Plausible Futures Using the World Green-HDI Measure

Table B.5

Safety Value Strategy and Milestone Strategies to Which It Is Compared

Safety Valve and Goals Strategies	Goal_N	Goal_S	Limit$_N$	Limit$_S$	T$_{init,N}$	T$_{init,S}$
Safety Valve	0%	0%	0.01	0.01	2010	2010
No Increase	0%	0%	none	none	2010	2010
M01	0%	1%	none	none	2010	2010
M02	0%	2%	none	none	2010	2010
M04	0%	none	none	none	2010	2010
M10	1%	0%	none	none	2010	2010
M11	1%	1%	none	none	2010	2010
Best Alternative	1%	2%	none	none	2010	2010
M14	1%	none	none	none	2010	2010
M20	2%	0%	none	none	2010	2010
M21	2%	1%	none	none	2010	2010
M22	2%	2%	none	none	2010	2010
M24	2%	none	none	none	2010	2010
M40	none	0%	none	none	2010	2010
M41	none	1%	none	none	2010	2010
M42	none	2%	none	none	2010	2010
Stay the Course	none	none	none	none	2010	2010

sample of 5,000 cases, which yielded 2,278 cases consistent with our constraints on the range of plausible population levels in 2050. The case that produced the highest regret for Safety Valve is shown in Table B.4. As with No Increase, this largest regret occurs for the World Green-HDI measure.

Figure 6.2 showed the results of the global sensitivity analysis for Safety Valve's regret. The most important input parameters are italicized in Table B.4. In contrast to No Increase, Safety Valve was relatively less sensitive to economic uncertainties and relatively more sensitive to the extent of the Southern sustainability challenge and the capabilities of future decisionmakers. We used these results to guide our choice of axes for the visualization in Figure 5.14.

It is also useful to examine the best performing strategy in those cases where Safety Valve fails. Figure B.2 shows the best strategy over the same set of futures shown in Figure 5.14. Note that in those futures where Safety Valve does worst the optimal response is generally Stay the Course or a similar strategy.

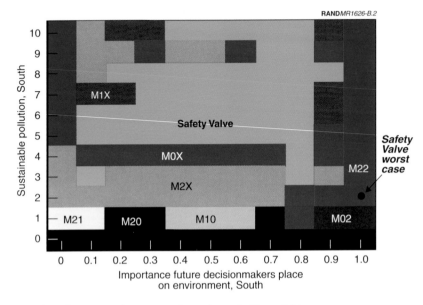

NOTE: These futures are the same as those shown in Figure 5.14.

Figure B.2—Optimum Strategy over a Landscape of Plausible Futures

BIBLIOGRAPHY

Aldis, Brian W., *Trillion Year Spree*, New York: Atheneum, 1986.

Allison, Graham, and Philip Zelikow, *Essence of Decision: Explaining the Cuban Missile Crisis*, second edition, Boston: Addison-Wesley, 1999.

Alkon, Paul K., *Origins of Futuristic Fiction*, Athens, Ga.: University of Georgia Press, 1987.

Azar, Christian, John Holmberg, and Sten Karlsson, *Decoupling— Past Trends and Prospects for the Future*, Stockholm, Sweden: Environmental Advisory Council, Ministry of the Environment, 2002.

Bain, David H., *Empire Express: Building the First Transcontinental Railroad*, New York: Penguin Books, 1999.

Bankes, Steven C., "Exploratory Modeling for Policy Analysis," *Operations Research*, Vol. 41, No. 3, 1993, pp. 435–449.

_____, "Agent-Based Modeling: A Revolution?" *Proceedings of the National Academy of Sciences*, Vol. 99, Supplement 3, 2002a, pp. 7199–7200.

_____, "Tools and Techniques for Developing Policies for Complex and Uncertain Systems," *Proceedings of the National Academy of Sciences*, Vol. 99, Supplement 3, 2002b, pp. 7263–7266.

Bankes, Steven C., and James Gillogly, "Validation of Exploratory Modeling," in Adrian Tentner, ed., *High Performance Computing:*

Grand Challenges in Computer Simulation, San Diego, Calif.: The Society for Computer Simulation, 1994a.

_____, *Exploratory Modeling: Search Through Spaces of Computational Experiments*, Third Annual Conference on Evolutionary Programming, World Scientific, 1994b.

Bankes, Steven C., Robert J. Lempert, and Steven W. Popper, "Computer-Assisted Reasoning," *Computing in Science and Engineering*, 2001, pp. 71–77.

Bankes, Steven C., Steven W. Popper, and Robert J. Lempert, *Incorporating Adversarial Reasoning in COAA Planning*, Topanga, Calif.: Evolving Logic, 2000.

Ben-Haim, Yakov, *Information-Gap Decision Theory: Decisions Under Severe Uncertainty*, Burlington, Mass.: Academic Press, 2001.

Benkard, C. Lanier, *Learning and Forgetting: The Dynamics of Aircraft*, Stanford, Calif.: Stanford Graduate School of Business, 1999.

Berger, James O., *Statistical Decision Theory and Bayesian Analysis*, New York: Springer-Verlag, 1985.

Brecke, Peter, "Integrated Global Models that Run on Personal Computers," *Simulation*, Vol. 60, No. 2, 1993.

Breiman, L., "Bagging Predictors," *Machine Learning*, Vol. 26, No. 2, 1996, pp. 123–140.

Brooks, Arthur, Steven Bankes, and Bart Bennett, "An Application of Exploratory Analysis: The Weapon Mix Problem," *Military Operations Research*, Vol. 4, No. 1, 1999.

Carley, Kathleen, "Computational Organizational Science: A New Frontier," *Proceedings of the National Academy of Sciences*, Vol. 99, supplement 3, 2002, pp. 7257–7262.

Cederman, Lars-Erik, *Emergent Actors in World Politics: How States and Nations Develop and Dissolve*, Princeton, N.J.: Princeton University Press, 1997.

Chamberlin, Thomas C, "The Method of Multiple Working Hypotheses," *Science*, 1890.

Cole, H. S. D., Christopher Freeman, Marie Jahoda, and K. L. Pavitt, eds., *Models of Doom: A Critique of the Limits to Growth*, New York: Universe Books, 1973.

Cropper, Mauren, and David Laibson, *The Implications of Hyperbolic Discounting for Project Evaluation*, World Bank Development Research Group, 1998.

Dalkey, N. C., and O. Helmer-Hirschberg, *An Experimental Application of the Delphi Method to the Use of Experts*, Santa Monica, Calif.: RAND, RM-727-PR, 1962.

Davis, Paul K., "Exploratory Analysis and Implications for Modeling," in Stuart E. Johnson et al., *New Challenges, New Tools for Defense Decisionmaking*, Santa Monica, Calif.: RAND, MR-1576-RC, 2003, pp. 255–283.

Davis, Steven, *The Diffusion of Process Innovations*, Cambridge, UK: The Cambridge University Press, 1979.

Dawes, Robyn M., *Rational Choice in an Uncertain World*, Orlando, Fla.: Harcourt Brace College Publishers, 1998.

DeCanio, Steve J., Catherine Dibble, and Keyvan Amir-Atefi, "The Importance of Organizational Structure for the Adoption of Innovations," *Management Science*, Vol. 46, No. 10, 2000, pp. 1285–1299.

_____, "Organizational Structure and the Behavior of Firms: Implications for Integrated Assessment," *Climate Change*, Vol. 48, 2001, pp. 487–514.

Dewar, James A., *The Information Age and the Printing Press: Looking Backward to See Ahead*, Santa Monica, Calif.: RAND, P-8014, 1998.

_____, *Assumption-Based Planning: A Tool for Reducing Avoidable Surprises*, Cambridge, UK: Cambridge University Press, 2001.

Dewar, James A., Steven C. Bankes, Sean J. A. Edwards, and James C. Wendt, *Expandability of the 21st Century Army*, Santa Monica, Calif.: RAND, MR-1190-A, 2000.

Dewar, James A., Steven C. Bankes, James S. Hodges, Thomas W. Lucas, Desmond K. Saunders-Newton, and Patrick Vye, *Credible Uses of the Distributed Interactive Simulation (DIS) Environment*, Santa Monica, Calif.: RAND, MR-607-A, 1996.

Dewar, James A., Carl H. Builder, William M. Hix, and Morlie H. Levin, *Assumption-Based Planning: A Tool for Very Uncertain Times*, Santa Monica, Calif.: RAND, MR-114-A, 1993.

de Cooman, G., T. L. Fine, and T. Seidenfeld, *Proceedings of the Second International Symposium on Imprecise Probabilities and Their Applications*, The Netherlands: Shaker Publishing, 2001.

Easterbrook, Gregg, *A Moment on the Earth: The Coming of Age of Environmental Optimism*, Viking Press, 1995.

Ellsberg, Daniel, "Risk, Ambiguity, and the Savage Axioms," *Quarterly Journal of Economics*, Vol. 75, 1961, pp. 644–661.

Epstein, Joshua, and Robert L. Axtell, *Growing Artificial Societies: Social Science from the Bottom Up*, Cambridge, Mass.: MIT Press, 1996.

Fonkych, Kateryna, *Modeling for Long-Term Policy Analysis: The Case of World3 Model*, RAND Graduate School: 15, 2001.

Forrester, Jay W., "Learning Through System Dynamics as Preparation for the 21st Century, Systems Thinking and Dynamic Modeling for K–12 Education," Concord Academy, 1994.

Gallopin, Gilberto, Al Hammond, Paul Raskin, and Rob Swart, *Branch Points: Global Scenarios and Human Choice*, Stockholm, Sweden: Stockholm Environmental Institute, 1997.

Gigerenzer, G., and P. M. Todd, *Simple Heuristics That Make Us Smart*, New York: Oxford University Press, 1999.

Gordon, T. J., and Olaf Helmer, *Report on a Long-Range Forecasting Study*, Santa Monica, Calif.: RAND, P-2982, 1964.

Gupta, S. K., and J. Rosenhead, "Robustness in Sequential Investment Decisions," *Management Science,* Vol. 15, No. 2, 1972 pp. 18–29.

Haeckel, S., and Adrian Slywotzky, *Adaptive Enterprise: Creating and Leading Sense-and-Respond Organizations,* Harvard Business School, 1999.

Hammitt, J. K., R. J. Lempert, and M. E. Schlesinger, "A Sequential-Decision Strategy for Abating Climate Change," *Nature,* Vol. 357, 1992, pp. 315–318.

Herbert, Ric. D., and Gareth D. Leeves, "Troubles in Wonderland," *Complexity International,* Vol. 6, 1998.

Hodges, J., "Six (or so) Things You Can Do with a Bad Model," *Operations Research,* Vol. 39, 1991, pp. 355–365.

Holmes, Stephen, *Passions and Constraints: On the Theory of Liberal Democracy,* Chicago: University of Chicago Press, 1995.

Hughes, Barry B., *International Futures: Choices in the Face of Uncertainty (Dilemmas in World Politics),* third edition, Boulder, Colo.: Westview Press, 1999.

Human Development Report, Oxford: Oxford University Press, 1990.

Human Development Report 2001: Making New Technologies Work for Development, Oxford: Oxford University Press, 2001.

ICSU, *Resilience and Sustainable Development,* International Council for Science, 2002.

Iman, R. L., J. M. Davenport, and D. K. Ziegler, *Latin Hypercube Sampling,* Sandia National Laboratory, 1980.

Kahn, Herman, William Brown, and Leon Martel, *The Next 200 Years—A Scenario for America and the World,* New York: William Morrow and Company, Inc., 1976.

Kahneman, Daniel, Paul Slovic, and Amos Tversky, *Judgment Under Uncertainty: Heuristics and Biases,* New York: Cambridge University Press, 1982.

Kahneman, Daniel, and Amos Tversky, *The Simulation Heuristic,* New York: Cambridge University Press, 1982.

Keeney, Ralph L. and Howard Raiffa, *Decisions with Multiple Objectives: Preferences and Value Tradeoffs,* New York: Wiley and Sons, 1976.

Kennan, George, "The Sources of Soviet Conduct," *Foreign Affairs,* 1947.

Klein, Gary, *Sources of Power: How People Make Decisions,* Cambridge, Mass.: MIT Press, 1998.

Knight, Frank H., *Risk, Uncertainty, and Profit,* Boston: Houghton Mifflin, 1921.

Kouvelis, P., and G. Yu, *Robust Discrete Optimization and Its Applications,* Dordrecht, the Netherlands: Kluwer Academic Publishers, 1997.

Kuhn, Thomas S., *The Structure of Scientific Revolutions,* Chicago: University of Chicago Press, 1962.

Lakátos, Imre, *Proofs and Refutations,* Cambridge, UK: Cambridge University Press, 1976.

Lempert, Robert J., "Finding Transatlantic Common Ground on Climate Change," *International Spectator,* Vol. 36, No. 2, 2001.

Lempert, Robert. J., "New Decision Sciences for Complex Systems," *Proceedings of the National Academy of Sciences,* Vol. 99, supplement 3, 2002a, pp. 7309–7313.

Lempert, Robert. J., *Transition Paths to a New Era of Green Industry: Technological and Policy Implications,* Santa Monica, Calif.: RAND, P-8067, 2002b.

Lempert, Robert J., and James Bonomo, *New Methods for Robust Science and Technology Planning,* Santa Monica, Calif.: RAND, DB-238-DARPA, 1998.

Lempert, Robert J., Steven W. Popper, and Steven C. Bankes, "Confronting Surprise," *Social Science Computing Review,* Vol. 20, No. 4, 2002, pp. 420–440.

Lempert, Robert J., Steven W. Popper, Susan Resetar, and S. Hart, *Capital Cycles and the Timing of Climate Change Mitigation Policy,* Arlington, Va.: Pew Center on Global Climate Change, 2002.

Lempert, Robert J., and Michael E. Schlesinger, "Robust Strategies for Abating Climate Change," *Climate Change,* Vol. 45, Nos. 3/4, 2000, pp. 387–401.

_____, *Adaptive Strategies for Climate Change. Innovative Energy Strategies for CO2 Stabilization,* R. Watts, ed., Cambridge, UK: Cambridge University Press, 2002.

Lempert, Robert J., Michael E. Schlesinger, Steven C. Bankes, and Natalie G. Andronova, "The Impact of Variability on Near-Term Climate-Change Policy Choices," *Climate Change,* Vol. 45, No. 1, 2000, pp. 129–161.

Leopold, Aldo, *A Sand County Almanac,* New York: Oxford University Press, 1949.

Lindblom, C. E., "The Science of Muddling Through," *Public Administration Review,* Vol. 19, 1959, pp. 79–88.

Lomborg, Bjorn, *The Skeptical Environmentalist: Measuring the Real State of the World,* Cambridge University Press, 2001.

Manne, A. S., and R. G. Richels, *Buying Greenhouse Insurance: The Economic Costs of Carbon Dioxide Emissions Limits,* Cambridge, Mass.: MIT Press, 1992.

Marland, Gregg, Tom Boden, and Bob Andres, "National CO2 Emissions from Fossil-Fuel Burning, Cement Manufacture, and Gas Flaring," 2002, available at http://cdiac.esd.ornl.gov/ftp/trends/emissions/usa.dat and http://cdiac.esd.ornl.gov/trends/emis/em_cont.htm.

Martin, Ben R., and John Irvine, *Research Foresight,* London: Pinter Publishers, 1989.

McKibbon, Bill, *The End of Nature,* New York: Random House, 1989.

McMillan, John, *Reinventing the Bazaar: A Natural History of Markets,* New York: W. W. Norton & Company, 2002.

Meadows, Donella H., and Dennis Meadows, eds., *The Limits to Growth: A Report for the Club of Rome's Project on the Predicament of Mankind*, second edition, New York: Universe Books, 1972.

Meadows, Donella H., Dennis L. Meadows, and Jorgen Randers, *Beyond the Limits: Confronting Global Collapse, Envisioning a Sustainable Future*, White River Junction, Vt.: Chelsea Green Publishing Co., 1992.

Metz, Bert, Ogunlade Davidson, Rob Swart, and Jiahua Pan, eds., *Climate Change 2001: Mitigation*, Contribution of Working Group III to the Third Assessment Report [TAR] of the Intergovernmental Panel on Climate Change (IPCC), Cambridge, UK: Cambridge University Press, 2001.

Morgan, Millett G., and Max Henrion, eds., *Uncertainty: A Guide to Dealing with Uncertainty in Quantitative Risk and Policy Analysis*, Cambridge, UK: Cambridge University Press, 1990.

Morgan, M. G., Milind Kandlikar, James Risebey, and Hadi Dowlatabadi, "Why Conventional Tools for Policy Analysis Are Often Inadequate for Problems of Global Change," *Climatic Change*, Vol. 41, 1999, pp. 271–281.

Myers, Norman, Nancy J. Myers, and Julian Simon, *Scarcity or Abundance?: A Debate on the Environment*, New York: W. W. Norton & Company, 1994.

Nakicenovic, N., *IPCC Special Report on Emissions Scenarios*, Cambridge, UK: Cambridge University Press, 1999.

National Academy Press, *Abrupt Climate Change: Inevitable Surprises*, Washington, D.C., 2002.

National Institute of Science and Technology Policy (NISTEP) and Fraunhofer Institute for Systems and Innovation Research, *Outlook for Japanese and German Future Technology*, NISTEP Report No. 33, April 1994.

Nature, "Climate Panel to Seize Political Hot Potatoes," Vol 421, p. 879, 2003.

Newell, Richard, and William Pizer, *Discounting the Benefits of Future Climate Change Mitigation: How Much Do Uncertain Rates Increase Valuations?* Arlington, Va.: Pew Center on Global Climate Change, 2001.

Nordhaus, W. D., *Managing the Global Commons: The Economics of Global Change,* Cambridge, Mass.: MIT Press, 1994.

Nordhaus, W. D., and Edward C. Kokkelenberg, eds., *Nature's Numbers: Expanding the National Economic Accounts to Include the Environment,* National Research Council, 1999.

Park, George, and Robert J. Lempert, *The Class of 2014: Preserving Access to California Higher Education,* Santa Monica, Calif.: RAND, MR-971-EDU, 1998.

Payne, John. W., James R. Bettman, and Eric J. Johnson, *The Adaptive Decision-Maker,* New York: Cambridge University Press, 1993.

Popper, Karl A., *The Logic of Scientific Discovery,* London: Hutchinson, 1959.

_____, *Conjectures and Refutations,* New York: Basic Books, 1962.

Popper, Steven W., Caroline Wagner, and Robert Lempert, "Moving Beyond Foresight," Foresight-Scenarios-Landscaping Workshop, Brussels, Belgium, July 5, 2003.

Prescott-Allen, Robert, *The Wellbeing of Nations,* Washington, D.C.: Island Press, 2001.

Raiffa, Howard, *Decision Analysis: Introductory Lectures on Choices Under Uncertainty,* Reading, Mass.: Addison-Wesley, 1968.

Raskin, Paul, Tariq Banuri, Gilberto Gallopin, Pablo Gutman, Al Hammond, Robert Kates, and Rob Swart, *Great Transition—The Promise and Lure of the Times Ahead,* Stockholm, Sweden: Stockholm Environment Institute, 2002.

Raskin, Paul, Gilberto Gallopin, Pablo Gutman, Al Hammond, and Rob Swart, *Bending the Curve: Toward Global Sustainability,* Stockholm, Sweden: Stockholm Environment Institute, 1998.

Resilience and Sustainable Development, International Council for Science, 2002.

Romer, C. D., "The Prewar Business Cycle Reconsidered: New Estimates of Gross National Product, 1969–1908," *Journal of Political Economy,* Vol. 98, No. 1, pp. 1–37.

Rosenhead, M. J., "Robustness Analysis: Keeping Your Options Open," in *Rational Analysis for a Problematic World: Problem Structuring Methods for Complexity, Uncertainty, and Conflict,* New York: Wiley and Sons, 1989.

Rosenhead, M. J., M. Elton, and S. K. Gupta, "Robustness and Optimality as Criteria for Strategic Decisions," *Operational Research Quarterly,* Vol. 23, No. 4, 1972, pp. 413–430.

Saltelli, A., K. Chan, and M. Scott, *Sensitivity Analysis,* New York: Wiley, 2000.

Savage, L. J., *The Foundations of Statistics,* New York: Wiley, 1950.

Schelling, Thomas, *Micromotives and Macrobehavior,* New York: W. W. Norton and Company, 1978.

Schwartz, Peter, *Art of the Long View,* New York: Doubleday, 1996.

Simon, Herbert, "Theories of Decision-Making in Economic and Behavioral Science," *American Economic Review,* 1959.

Sobol, I. M., "Sensitivity Estimates for Nonlinear Mathematical Models," *Matematicheskoe Modelirovanie,* Vol. 2, 1990, pp. 112–118.

Szayna, Thomas S., Daniel Byman, Steven Bankes, Derek Eaton, Seth G. Jones, Robert Mullins, Ian O. Lesser, and William Rosenau, eds., *The Emergence of Peer Competitors: A Framework for Analysis,* Santa Monica, Calif.: RAND, MR-1346-A, 2001.

Trigeorgis, L., *Real Options: Managerial Flexibility and Strategy in Resource Allocation,* Cambridge, Mass.: MIT Press, 1996.

United Nations Development Programme (UNDP), *Human Development Report,* Oxford, UK: Oxford University Press, 1990.

_____, *Human Development Report 2001: Making New Technologies Work for Development,* New York: Oxford University Press, 2001.

Utterback, James M., *Mastering the Dynamics of Innovation,* Boston: Harvard Business School Press, 1994.

van Asselt, Marjolein B. A., *Perspectives on Uncertainty and Risk,* Dordrecht, the Netherlands: Kluwer Academic Publishers, 2000.

van der Heijden, K., *Scenarios: The Art of Strategic Conversation,* Chichester, UK: Wiley and Sons, 1996.

Victor, David, *The Collapse of the Kyoto Protocol and the Struggle to Slow Global Warming,* Princeton, N.J.: Princeton University Press, 2001.

Vitousek, Peter, Paul R. Ehrlich, Anne H. Ehrlich, and Pamela Matson, "Human Appropriation of the Products of Photosynthesis," *BioScience,* 1986.

Walter, Dave, ed., *Today Then: America's Best Minds Look 100 Years into the Future on the Occasion of the 1893 World's Columbian Exposition,* Helena, Mont.: American & World Geographic Publishing, 1992.

Walters, Carl, *Adaptive Management of Renewable Resources,* Caldwell, N.J.: The Blackburn Press, 1986.

Watson, Stephen R., and Dennis M. Buede, *Decision Synthesis,* Cambridge, UK: Cambridge University Press, 1987.

Weitzman, Martin L., "Why the Far-Distant Future Should Be Discounted at Its Lowest Possible Rate," *Journal of Environmental Economics and Management,* Vol. 36, 1998, pp. 201–208.

Zhou, K., J. Doyle, and K. Glover, *Robust and Optimum Control Theory,* Englewood Cliffs, N.J.: Prentice-Hall, 1996.